Mental Maps

The concept of mental maps is used in several disciplines including geography, psychology, history, linguistics, economics, anthropology, political science, and computer game design. However, until now, there has been little communication between these disciplines and methodological schools involved in mental mapping.

Mental Maps: Geographical and Historical Perspectives addresses this situation by bringing together scholars from some of the related fields. Ute Schneider examines the development of German geographer Heinrich Schiffers' mental maps, using his books on Africa from the 1930s to the 1970s. Efrat Ben-Ze'ev and Chloé Yvroux investigate conceptions of Israel and Palestine, particularly the West Bank, held by French and Israeli students. By superimposing large numbers of sketch maps, Clarisse Didelon-Loiseau, Sophie de Ruffray, and Nicolas Lambert identify "soft" and "hard" macro-regions on the mental maps of geography students across the world. Janne Holmén investigates whether the Baltic and the Mediterranean Seas are seen as links or divisions between the countries that line their shores, according to the mental maps of high school seniors. Similarly, Dario Musolino maps regional preferences of Italian entrepreneurs. Finally, Lars-Erik Edlund offers an essayistic account of mental mapping, based on memories of maps in his own family.

This edited volume uses printed maps, survey data and hand drawn maps as sources, contributing to the study of human perception of space from the perspectives of different disciplines.

The chapters in this book were originally published as a special issue of the *Journal of Cultural Geography.*

Janne Holmén is Associate Professor of History of Education at Uppsala University, Sweden, and researcher at the Institute of Contemporary History, Södertörn University, Sweden. His research focuses on comparative studies in educational history, with an emphasis on the Nordic countries. He has published studies on mental maps and historical consciousness.

Norbert Götz is Professor at the Institute of Contemporary History, Södertörn University, Sweden. His research interests include spaces of civil society, humanitarianism, international relations, social transformations, populism, and moral economy.

Mental Maps

Geographical and Historical Perspectives

Edited by

Janne Holmén and Norbert Götz

Routledge
Taylor & Francis Group
LONDON AND NEW YORK

First published 2022
by Routledge
2 Park Square, Milton Park, Abingdon, Oxon, OX14 4RN

and by Routledge
605 Third Avenue, New York, NY 10158

Routledge is an imprint of the Taylor & Francis Group, an informa business

British Library Cataloguing-in-Publication Data
A catalogue record for this book is available from the British Library

ISBN13: 978-1-032-11440-8 (hbk)
ISBN13: 978-1-032-11441-5 (pbk)
ISBN13: 978-1-003-21994-1 (ebk)

DOI: 10.4324/9781003219941

Typeset in Minion Pro
by codeMantra

Publisher's Note
The publisher accepts responsibility for any inconsistencies that may have arisen during the conversion of this book from journal articles to book chapters, namely the inclusion of journal terminology.

Contents

Citation Information

The chapters in this book were originally published in the *Journal of Cultural Geography*, volume 35, issue 2 (2018). When citing this material, please use the original page numbering for each article, as follows:

Introduction

Introduction to the theme issue: "Mental maps: geographical and historical perspectives"
Norbert Götz and Janne Holmén
Journal of Cultural Geography, volume 35, issue 2 (2018) pp. 157–161

Chapter 1

Dimensions of remapping: Heinrich Schiffers and his mental map of Africa
Ute Schneider
Journal of Cultural Geography, volume 35, issue 2 (2018) pp. 162–188

Chapter 2

Palestine, Israel, Gaza, and the West Bank: the muddled mental maps of French and Israeli students
Efrat Ben-Ze'ev and Chloé Yvroux
Journal of Cultural Geography, volume 35, issue 2 (2018) pp. 189–209

Chapter 3

Mental maps of global regions: identifying and characterizing "hard" and "soft" regions
Clarisse Didelon-Loiseau, Sophie de Ruffray and Nicolas Lambert
Journal of Cultural Geography, volume 35, issue 2 (2018) pp. 210–229

Chapter 4

Changing mental maps of the Baltic Sea and Mediterranean regions
Janne Holmén
Journal of Cultural Geography, volume 35, issue 2 (2018) pp. 230–250

Chapter 5

The mental maps of Italian entrepreneurs: a quali-quantitative approach
Dario Musolino
Journal of Cultural Geography, volume 35, issue 2 (2018) pp. 251–273

Chapter 6

Creative Mappings: Some reflections on mental maps
Lars-Erik Edlund
Journal of Cultural Geography, volume 35, issue 2 (2018) pp. 274–285

For any permission-related enquiries please visit:
http://www.tandfonline.com/page/help/permissions

Notes on Contributors

Efrat Ben-Ze'ev, The Department of Behavioral Sciences, The Ruppin Academic Center, Emek Hefer, Israel.

Clarisse Didelon-Loiseau, UFR 08 Géographie, Université Paris 1 Panthéon Sorbonne, UMR Géographie-cités, Paris, France.

Lars-Erik Edlund, Department of Language Studies, Umeå University, Sweden.

Norbert Götz, Institute of Contemporary History, Södertörn University, Huddinge, Sweden.

Janne Holmén, Institute of Contemporary History, Södertörn University, Huddinge, Sweden; Department of Education, Uppsala University, Sweden.

Nicolas Lambert, UMS RIATE, CNRS, Paris, France.

Dario Musolino, Bocconi University, Milan, Italy; Department of Economics and Political Science, Università della Valle d'Aosta, Aosta, Italy.

Sophie de Ruffray, Département de géographie, Université de Rouen, UMR IDEES, France.

Ute Schneider, Department of History, Social and Economic History, University of Duisburg-Essen, Germany.

Chloé Yvroux, LAGAM Department of Geography, Paul Valery University of Montpellier, France.

Introduction: "Mental maps: geographical and historical perspectives"

Norbert Götz and Janne Holmén

Maps are symbolic representations of spatial features. As such, they are by definition projections that involve choices of inclusion and modes of depiction. They are therefore subject to framing, coding, and graphic design in their conception and execution. Even the most positivist attempt to map the world as it is – to represent a set of properties systematically, scaling the matrix in which they are embedded in proportion to their ratio in the physical environment – involves a mental conversion. Thus, all maps, from those on classroom walls, to fold-outs of city streets and subways, to GPS on smartphones and aircraft screens, are "mental maps" whose design rests on the decisions of mapmakers.

However, although all maps are artistic conceptions, a fundamental distinction may be drawn between maps proper – that is, those that are fixed cartographic manifestations of spatial relations – and mental maps, whose spatialization of meaning dwells latently in the minds of individuals or groups of people. Visually realized maps can be analyzed to give insight into the underlying mental maps that have shaped them, laying bare mindsets or agendas that may be as socio-culturally significant as the geography they present. In addition, they may often contain prescriptive images that incidentally shape mental maps in those who view them, thus implanting or concretizing social knowledge. In this issue, Ute Schneider examines the development of German geographer Heinrich Schiffers' mental maps with reference to cartographic illustrations in his books on Africa from the 1930s to the 1970s.

Mental maps can also be decoded to reveal biases of objectified cartographic knowledge such as socio-spatial hierarchies that structure the world, or to explore ways in which collectives and individuals orient themselves in their environment, or to understand how they perceive the world. One way of elucidating mental maps is by examining hand-drawn sketches by informants of various backgrounds. In one of our articles, Efrat Ben-Ze'ev and Chloé Yvroux use this method to disclose conceptions of Israel and Palestine, particularly the West Bank, held by French and Israeli students. In another, by superimposing large numbers of sketch maps, Clarisse Didelon-Loiseau, Sophie de Ruffray, and Nicolas Lambert identify "soft" and "hard" macroregions on the mental maps of geography students across the world. A deliberately personal approach to mental mapping is represented in this issue by Lars-Erik Edlund, who offers an essayistic account of mental mapping from a liberal arts perspective, taking as a point of departure memories of maps in his own family.

Although many researchers call the subjective map-drawings of their informants mental maps, implying a distinction between fictitious mental maps and their real counterparts, we prefer a more formal distinction between *charted maps* (endowed with varied claims of objectivity), and latent *mental maps* (with correlations to the physical world). As we see it, a mental map, rather than being an object, is a theoretical construct not observable in its original repository – the human brain. It is accessible to scrutiny only when reified via behavioral, oral, textual, or graphical acts.

However, the meta-perspective taken by investigations of mental maps complicates the picture. Researchers frequently summarize their findings with regard to mental maps of certain populations through cartographic illustrations. Such images are neither latent in the minds of people, nor are they firsthand, pre-analytic representations of spatial knowledge. They qualify as mental maps because they graphically articulate conceptual notions of space. Thus, researchers are not concerned with the utility of such maps for transversing space, but rather wish to understand the contingent apperception of the world contained in those maps. Janne Holmén charts such mental maps, investigating whether the Baltic and the Mediterranean Seas are seen as links or divisions between the countries that line their shores, according to the mental maps of high school seniors. Similarly, Dario Musolino reconstructs mental maps in order to understand regional preferences of Italian entrepreneurs.

Alongside "cognitive map", an approximate synonym with more neurological connotations, the concept of mental map is well established in geography, behavioral science, and psychology. Immanuel Kant may have anticipated the idea of mental maps in his writings on geography (Richards 1974), but it was in the interwar period that psychologist Edward Tolman developed a modern understanding of spatial orientation, and later coined the term cognitive map in his studies of learning in rats (1948). From the 1960s onwards, behavioral geographers came to develop a related interest in the depiction of space in the human mind. Although not actually utilizing the term mental map, Kevin Lynch's book *The Image of the City* (1960) is regarded as a pioneering work in the field. Another key figure was Peter

Gould, who called his isolinear maps of how people perceived different areas mental maps (Gould 1966; Gould and White 1974).

Influential contributors to the discussion of mental maps from other disciplines did not explicitly use the concept themselves. Benedict Anderson's *Imagined Communities* (1983) and Edward Said's *Orientalism* (1979) are probably the foremost works to have advanced the concept of mental maps in the humanities and social sciences. They examine two opposed socio-cultural processes, communitization and "othering", that have seen particular interest in mental mapping research.

Historians have also taken up the idea of mental maps over the past 25 years, especially in analyzing patterns of dominance and subalternity, the construction and dissolution of historical regions, and the world views of political elites. The German journal *Geschichte und Gesellschaft* was comparatively early in publishing a special issue on the topic (Conrad 2002), and despite the promotion of such competing concepts as environmental images or spatial representation, that of mental maps "has become fairly standard in historical research on collective concepts of geographical and historical macro-regions" (Schenk 2013, see, e.g. Mishkova forthcoming).

Disciplines that have found the concept of mental maps useful include geography, psychology, history, linguistics, economics, anthropology, political science, and computer game design. To date, there has been little communication between those disciplines and methodological schools involved in mental mapping, and an international multi-disciplinary conversation on mental maps with an emphasis on cultural patterns is still in its early stages.[1] This special issue of the *Journal of Cultural Geography* addresses this situation by bringing together scholars from the fields of history, geography, economics, anthropology, and linguistics, and by including a variety of quantitative and qualitative research methods. The authors presented here are affiliated with research institutions in Finland, France, Germany, Israel, Italy, and Sweden. They show that mental mapping research is an exciting arena for inter-disciplinary and international encounters. We believe their fascinating accounts also demonstrate the potential for the further expansion of the field.

The idea of this themed issue emerged from a workshop entitled "Mental Mapping – Historical and Social Science Perspectives" held on 12–13 November 2015 at the Institute of Contemporary History, Södertörn University, and the Italian Cultural Institute "C.M. Lerici" in Stockholm. The keynote speaker was Larry Wolff, whose account of the development of the mental map of Eastern Europe from the Enlightenment to Harry Potter is published separately (Wolff 2016). Other contributors included Jonathan Wright, editor of three volumes on the mental maps of leading politicians in the twentieth century (Casey and Wright 2008, 2011, 2015), and Thomas Scheffler, who has studied conflicting mental maps of Lebanon and the Middle East

(Scheffler 2003). The workshop was arranged by the research project "Spaces of Expectation: Mental Mapping and Historical Imagination in the Baltic Sea and Mediterranean Region", a joint venture between Södertörn University and Ca' Foscari University in Venice that is funded by the Foundation for Baltic and East European Studies.

The project has published other studies related to those presented here in the special issue "Maritime Areas: Spaces of Changing Expectations" in the journal *Comparativ* (Petri 2016a), as well as articles on the cohesion of regions (Petrogiannis and Rabe 2016), on the Mediterranean metaphor in geopolitics (Petri 2016b), on the use of mental mapping techniques for surveying historical consciousness (Holmén 2017), and on the fuzzy topography of international organization of the Baltic Sea region (Götz 2016).

In the current era of disoriented globalization, we believe mental maps will continue to be crucial tools for insights into the ability of ordinary people to make sense of the world and into the compasses of their political leaders. In addition, mental mapping may contribute to an improved understanding of the effects of multiple spatial frames conveyed by political institutions and various social organizations (Götz 2008), including attempts at place branding (Gertner 2011), efforts to create areas of limited statehood (Risse 2011), and other forms of manipulating space. Finally, despite the uneasiness about the era we are living in, mental maps show that any juxtaposition of "post-truth" and truth fails to do justice to the ineluctable subjectivity of the human condition.

Funding

This work was supported by the Foundation for Baltic and East European Studies [grant number 41/13].

Note

1. See, however, the cross-disciplinary discussion in the German anthology *Die Ordnung des Raums: Mentale Landkarten in der Ostseeregion* (The order of space: Mental maps in the Baltic Sea region; Götz *et al.* 2006).

ORCID

Norbert Götz http://orcid.org/0000-0002-8788-101X
Janne Holmén http://orcid.org/0000-0003-2449-4888

References

Anderson, B., 1983. *Imagined communities: reflections on the origin and spread of nationalism*. London: Verso.

Casey, S. and Wright, J., eds., 2008. *Mental maps in the era of two world wars.* Basingstoke: Palgrave Macmillan.
Casey, S. and Wright, J., eds., 2011. *Mental maps in the era of the early cold war 1945-68.* Basingstoke: Palgrave Macmillan.
Casey, S. and Wright, J., eds., 2015. *Mental maps in the era of détente and the end of the cold war 1968-91.* Basingstoke: Palgrave Macmillan.
Conrad, Ch., ed., 2002. Mental maps. *Geschichte und Gesellschaft*, 28 (3), 339–514.
Gertner, D., 2011. A (tentative) meta-analysis of the "place marketing" and "place branding" literature. *Journal of Brand Management*, 19 (2), 112–131.
Götz, N., 2008. Western Europeans and others: the making of Europe at the United Nations. *Alternatives*, 33 (3), 359–381.
Götz, N., 2016. Spatial politics and fuzzy regionalism: the case of the Baltic Sea area. *Baltic Worlds*, 9 (3), 54–67.
Götz, N., Hackmann, J., and Hecker-Stampehl, J., eds, 2006. *Die Ordnung des Raums: Mentale Landkarten in der Ostseeregion.* Berlin: Wissenschafts-Verlag.
Gould, P., 1966. *On mental maps.* Ann Arbor: University of Michigan Press.
Gould, P. and White, R., 1974. *Mental maps.* Harmondsworth: Pelican.
Holmén, J., 2017. Mapping historical consciousness: mental maps of time and space among secondary school students from ten locations around the Baltic and Mediterranean seas. *Autonomy and Security*, 1(1): 46–74.
Lynch, K., 1960. *The image of the city.* Cambridge, MA: Technology Press.
Mishkova, D., forthcoming. *Beyond Balkanism: the scholarly politics of mental maps.* Abingdon: Routledge.
Petri, R., ed., 2016a. Maritime areas: spaces of changing expectations. *Comparativ* [Special Issue], 26 (5), 7–75.
Petri, R., 2016b. The Mediterranean metaphor in early geopolitical writings. *History*, 101 (348), 671–691.
Petrogiannis, V. and Rabe, L., 2016. *What is it that holds a region together? Baltic Worlds* [In-house edition], 5–9.
Richards, P., 1974. Kant's geography and mental maps. *Transactions of the Institute of British Geographers*, 61, 1–16.
Risse, T., 2011. *Governance without a state? Policies and politics in areas of limited statehood.* New York, NY: Columbia University Press.
Said, E. W., 1979. *Orientalism.* New York, NY: Vintage.
Scheffler, T., 2003. "Fertile Crescent", "Orient", "Middle East": the changing mental maps of Southwest Asia. *European Review of History*, 10 (2), 253–272.
Schenk, F. B., 2013. *Mental maps: the cognitive mapping of the continent as an object of research of European history.* European History Online *(EGO)*. Available from: http://www.ieg-ego.eu/schenkf-2013-en [Accessed 16 April 2017].
Tolman, E.C., 1948. Cognitive maps in rats and men. *Psychological Review*, 55 (4), 189–208.
Wolff, L., 2016. *Mental mapping and Eastern Europe.* Huddinge: Södertörn University (= Södertörn Lectures 12).

Dimensions of remapping: Heinrich Schiffers and his mental map of Africa

Ute Schneider

ABSTRACT
At the beginning of the 1960s, the Athenäum publishing house in Germany planned a revised and extended edition of Heinrich Schiffers' (1901–1982) successful book *Wilder Erdteil Afrika* (English translation: *The Quest for Africa*). The bestselling author had published several monographs about Africa since the 1930s, and authored and edited numerous works after World War II. Nearly all of these works, whose substantial print runs are testament to their popularity, are characterized by an engaging combination of text, images, and cartographic material, creating narratives and mental maps about Africa, its history, and the colonial past. In his later writings, he stressed the importance of "relearning" with regard to Africa and struggled to remap the imaginative geography of Africa. In this paper, I examine the characteristics of Schiffers' imaginative geography and the change in his writings and maps. I explore whether his concept of "relearning" was an epistemological decolonization or if there were any continuities found in his imaginative geography. In order to grasp the specifics of his thinking, his geography will be briefly compared with that of his contemporary, Frankfurt zoo director Bernhard Grzimek.

Introduction

In 1977, linguist and Arabist Karl Stowasser (1977) reviewed a book by geographer Heinrich Schiffers in the renowned *Middle East Journal*. Schiffers had published a historical and political regional geography of Libya together with two colleagues under the title *Libya: Burning Desert – Blooming Sand*. Despite an overall positive review, Stowasser included the following remark about the style of the book: "Some readers may also feel slightly irritated by the jaunty and somewhat breathless 'gee-whiz-would you believe that?' (1977, p. 220) exclamation-point-ridden style in which a good deal of the material is presented." He mentioned in amusement the constant reminders given to readers to refrain from a Eurocentric view and smugly stated: "the index even contains a curious item '*umlernen*'" (1977, p. 220). This emphasis on "relearning" is noticeable insofar as it is not a technical term. Schiffers seems rather to be using it to describe a process of epistemological decolonization, which his thinking and geography had to pass through since his first books in the inter-war period had been published. However – and this makes Schiffers' use of "umlernen" quite interesting – it not only means an active process of relearning old episteme but includes echoes of the democratic re-education of the Germans after 1945.

A glance into the book confirms this impression. Right on the first pages, Schiffers (1975, pp. 9–12) paints a panoramic view of Libya as a country between tradition and modern age, between "pious Islamic sharifs" and "Rommel's tank brigades", between shepherds and oil traders. Since the discovery of oil at the end of the 1950s, the country experienced a boom and the "bold Libyan reform experiment" attracted business interest and Western attention. The "Westerners" however, argued Schiffers, not only had considerable knowledge gaps but held a view of Libya that originated from colonial times and focused on the coastal region. He demanded a rethink of the approach for the "project era" – as he named the process of decolonization and development policy – that had just begun. Moreover, he argued, the Europeans were the ones who had to do the relearning. They had to comprehend, for example, that their continent and the "Western" world with it, the "centre of all values and yardsticks" for centuries, "has not pointed the way for the Third (and Fourth[1]) World for quite some time now".

Heinrich Schiffers did not exclude himself from the necessity of this process. He and his two "travel companions" had made themselves familiar with the country and its people since their first immersion in the "Islamic-Arabic world zone" in 1958. As a "Westerner willing to relearn", he had had to completely revise his first publication from the same year just four years later. The resubmitted book was based on "long discussions", neither being intended to be a "journalistic portrait" nor a "regional fact book filled with geographic or ecological stereotypes and many tables" (1975, p. 11).

The fairly popular Libya book was among the last of Schiffers' works, which had been preceded by a large number of similarly popular publications that had seen repeated editions and reprints since 1935. Almost all of his publications found not only national and international attention; he was one of the more well-known authors of geographical handbooks after the Second World War. His career moved between science and school service and ended in the 1970s in the respectable and influential Munich Institute for Economic Research (IFO, today CESifo). Against the background of this extensive oeuvre and his influential position contributing fundamentally to the knowledge of Africa in the Federal Republic, his emphasis on "relearning" raises the question of what Schiffers' perceptions of Africa at the different periods of his life and the different social and political contexts were like.

Schiffers' continuous publication activities since the inter-war period provide a suitable basis for exploring the dimensions of his "relearning" from the texts, images, and maps. His emphasis on a conscious process of "relearning" raises the question as to how far imaginative geographies can be subjected to an intentional process of transformation, and whether an epistemological decolonization can be consciously accomplished. Such a perspective is directly related to research on "mental maps", which examines among other aspects the worldviews, values and models of social groups and individual actors. In particular, the concept of Orientalism has encouraged work on imaginative geographies and the mental models of influential actors (Kitchin and Freundschuh 2000; Power and Sidaway 2004; Andreasson 2005; Tilley 2011).

In contrast to British and French colonial officials, however, the question of Schiffers' "imaginative geography" and the meaning of "relearning" in his case still have an additional political dimension. After all, with the end of the "Third Reich" in 1945, all colonial ambitions and research projects, which had continued since 1918 ended abruptly. In spite of the loss of their colonies at the end of World War I, some German geographers continued their research on Africa as a colonial space after 1918 and co-operated after 1937 with the colonial-political office of the NSDAP (Schultz 1989; Haar and Fahlbusch 2005; Jureit 2012; Schneider 2012). After the war, Heinrich Schiffers was among the German researchers who had to undergo the formal de-nazification carried out by British officers and he, along with all Germans, was subjected to the process of "reeducation". "Reeducation" meant for the British and American Allies a guided and comprehensive process of democratization of the German population. The question then arises as to whether Schiffers' concept of "relearning" went beyond this democratization process and led to a change in his perception and imaginative geography of Africa. To some extent, I will argue, Schiffers succeeded in transforming his imaginative geography, but the epistemological decolonization had its limitations insofar as it remained attached to a European narrative of modernization and geopolitical thinking.

Past studies of mental maps have taken their evidence mainly from written sources, and have largely ignored visual evidence (Werlen 2008; Schenk 2013). Indeed, the inclusion of photo and, above all, map material has been given comparatively little attention, although the research on mental maps has pointed to the power of visualizations and the link between text and image. Beyond it, this matters because cognitive mapping research continues to debate the status of maps and their influence on spatial behavior, as well as the relation between reception and literacy, and finally the media practices of mapping (Gregory 1994; Kitchin and Blades 2001; Cosgrove 2008; Gieseking 2013).

This study works to bring the visual into the study of mental maps through an exploration of the evolving work of Schiffers. His writings allow such a combined text-image analysis, because his books always contained extensive visual material. An analysis of both also provides insights into the coherence or potential tensions between "relearning" and his imaginative geography on the verbal and visual levels. This is important not least because Schiffers' publications gained a widespread audience, and they have simultaneously exposed knowledge, world images, and representations. The following considerations will focus on Schiffers' work, including text, photos and maps, and look for evidence of his "relearning" his own imaginative geographies.

Career path between academia and politics

Heinrich Schiffers was born in 1901, the son of a master baker in Aachen, and attended a humanist grammar school until the lower sixth grade. In 1921, he passed his first teacher's exam and three years later in Cologne the supplementary exam in Latin and Greek that was required for full-fledged enrollment. He then studied German, History, and French. In 1940 and 1941, he enrolled as a guest student in Cologne, with geography as his main subject.[2] At the same time, he worked as a teacher of geography and French at various schools starting in 1927. He was drafted in 1941, and according to his own statements, worked as an interpreter in the military. His name is included on a list of "speakers essential to the war effort" by Hitler's "Beauftragter für die Überwachung der gesamten geistigen und weltanschaulichen Schulung und Erziehung der NSDAP", a kind of commissioner for supervising the entire mental and philosophical training and education of the NSDAP from 1940 and he was a member of the "Reichskolonialbund" (Reich Colonial League), a union of all colonial associations and organizations with the aim to regain colonies, until 1941.[3]

Schiffers found the time to write a dissertation after a first book manuscript titled "The Sahara in Europe's View", which was destroyed in a fire in 1943. In August 1944, he obtained his doctorate in Marburg under Helmuth Kanter (1891–1976) with research on Algeria titled "The Ténéré as a Type of North African Desert Region." Schiffers maintained a close relationship with his doctoral supervisor, a geographer and medical scientist, who was awarded a chair in Marburg because of his membership in the Nazi party. Kanter recruited several convinced national socialists during his professorship, and an interest in Africa and Libya linked him with his student, Schiffers.[4] Years later, Schiffers thanked Kanter in his preface to *Libya: Burning Desert* (1975, p. 12).

The focus of Schiffers' dissertation was an explanation of the nature of the Ténéré, a desert region in the South Sahara, "based on the region". The work clearly has an anti-French tenor, speaking out against "Europeans alien to the region" or "elements", as he calls them in some passages (1944, p. 136).

The maps Schiffers created of the region, form part of his dissertation. These maps unfortunately cannot be viewed because the text of the dissertation and the maps were separated during archiving and the maps got lost. The dissertation was not Schiffers' first work about Africa. He had published a travel report about North Africa titled "People under Allah's Sun", still under the name of Schiffers-Davringhausen in 1935. In 1936, this was followed by "Silent Front. Men and Powers under the Spell of the Sahara" (Schiffers-Davringhausen 1936). In this highly popular work, he for the first time, and more explicitly than in the earlier travel report, combined reference to travel experiences from the nineteenth century with contemporary African history, especially the conquest of Libya by Italian troops.

After World War II, Schiffers published – now no longer under his double name – geographical works about Africa and the Sahara, such as a volume of the Harms handbooks of geography, a brief regional geography of Africa (1953, 1962a; Schiffers et al. 1957; Schiffers and Wagner 1973). Other works included *Libya and the Sahara* (1962b), while others examined *Sahel zone: its future* (1976). While most of these works were published in several editions, his monographs compiling historical sources, especially travel reports of Africa, were much more popular and had a more lasting influence on the perception of Africa in the Federal Republic of Germany and later generations of geographers and historians. Among these were *The Big Journey*, about Heinrich Barth's adventures and research (Schiffers 1952), and Schiffers' comprehensive work about the "great expeditions" of European explorers in Africa, entitled *Africa, the Wild Continent* (1954, also published in an extended edition in 1962) (Schiffers 1954, Schiffers 1962c). Schiffers was not just hugely productive but also widely read. This is apparent from an English translation of *The Wild Continent* published in London (1957) and one year later by Putnam in New York (1958) under the title *The Quest for Africa*, and also from numerous reviews that most of his books received in international professional publications. German scholars also acknowledged Schiffers' works. The Leipzig historian Gerhard Jacob, editor of *Deutsche Kolonialpolitik in Dokumenten* (1938), highlighted the significance of the work for scholarship into colonial Africa:

> If one considers that German colonial history has been paid little attention by German historians in books and bibliographies, apart from a few exceptions […], this book, which deals with German research and colonial history as part of European African history, deserves special mentioning for this very reason alone. (Jacob 1956, p. 232; see also Hellen 1968, 1970, 1974; Maxwell 1959)

In view of the popularity of Schiffers' works, publisher Paul Junker of Athenäum approached the German Research Foundation (DFG) to elicit its interest in an African atlas.[5] Junker said that

> the intention of Dr. Schiffers, another edition of whose Africa book is going to be published soon, to create a great atlas of Africa with a number of co-workers and not only show the political and geographical conditions but also capture other issues of an anthropological and economic nature.[6]

By structuring the atlas in this way, Schiffers aimed the atlas not only at an interested public, but also towards development assistance professionals, particularly scientists and politicians responsible for the international program "AID" (Agency for International Development), which the Federal Republic of Germany had only adopted as an area of action a few years previously under pressure from the Western powers (Ansprenger 1967; Büschel 2014; Eckert 2015).

Against this backdrop of political scientific topicality, the DFG appeared very interested, and their specialist on the subject agreed with Schiffers on the phone that his organization was

> certainly willing to support a scientific project with a research grant and printing grants later. But it would really have to be scientific research, not a current political atlas of Africa, which every major daily newspaper prints about once in 6 months. [...] Dr. Schiffers will discuss the details with Dr. Manshard[7], whom he named himself, and formulate a draft application.[8]

In parallel, the representative used several rounds of informal talks with geographers to obtain information about Schiffers. He noted in a comment: "Dr. Schiffers was certainly known to the gentlemen. He is said to be about 60 years old and capable of performing good scientific work. He has produced quite remarkable publications."[9] However, although there were no fundamental misgivings about the substance of the project, the fact that Schiffers was not based in any institute and – holding the title of *Studienrat*, a senior teacher at secondary school level – did not have a scientific reputation, went against him, as the representative explained in his further reasoning:

> A new atlas of Africa would certainly be welcome. [...] Such an atlas would be a major undertaking in which a number of disciplines would have to be involved. It would need firm leadership. Such a work cannot be done without the support of an institute. But that means that Dr. Schiffers cannot lead the project. He will be able to make important contributions. But it will need greater scholars [sic] who work on special aspects, and a manager who pushes the project forward like an engine.[10]

Two years and numerous talks among the DFG and geography professors later, the DFG launched its *Atlas of Africa* Priority Program which provided jobs and career opportunities for generations of geographers (Schneider 2012). There was no more talk of involving the initiator, Schiffers, at this point; he was excluded from discussions of the project. Although his attempts to find favor with the scientific establishment remained unsuccessful, Schiffers' exclusion from the project did not in any way spell the end of his career. Since he had moved between science and politics as an author since the 1930s, he ended up at the Institute for Economic Research (IFO) in Munich, where he was able to develop and publish projects at the African Studies Centre. This group of scientists specializing in Africa within the IFO was financially supported by the Fritz Thyssen Foundation and had been presided over by the geographer Walter Marquardt since 1964.

However, there is a possibility that his excellent contacts with former party members and like-minded individuals eased his integration into the Munich-based institute, which had been presided over by Hans Langelütke (1892–1972) since 1955. The latter already knew his departmental head Marquardt from the time that they were both responsible for economic statistics in the planning office of the Nazi Four Year Plan. In a broader sense, Langelütke's successor must also be included in this network, since Karl Maria Hettlage (1902–1995) had managed one of the three main divisions of the General Construction Inspectorate after 1940 (Aly and Heim 1993, p. 54; Schrafstetter 2008, p. 431).

Even after the end of the "Third Reich", Schiffers had depended on protection by friends and colleagues, for his denazification process was by no means as smooth as he had hoped. He was classified as a sympathizer with National Socialism (Mitläufer), and only after his appeal in 1947 was he again admitted to school service. The geographer Carl Troll (1899–1975) of the University of Bonn had issued him one of the needed denazification certificates in 1945, because he was acquainted with Schiffers as an African explorer and scientist. "As far as the political problems of Africa are concerned, we always discussed sharply the colonial policy, as it was conceived and operated by the Reichskolonialbund and the colonial-political office of the NSDAP." [11] Troll, who had also dealt with colonial spatial planning in Africa at the beginning of the 1940s, successfully continued his career in the Federal Republic and internationally after 1945. He was one of the editors of the *Atlas of Africa* Priority Program, he decided on money for geographical projects and professional careers in the German Research Foundation (DFG), and he was president of the International Geographical Union (IGU) from 1960 to 1964. In 1947, he had addressed German and American geographers with an article in which he pursued a self-cleaning and relief strategy for German geography during National Socialism. This essay did not only influence the (self)-perception of German geographers but also promoted their successful cooperation with the Americans after 1945 (Smith and Black 1946; Troll 1947, 1949; Böhm 2003; Butzer 2004; Zimmerer 2004).

Even as a figure at the margins of professional geography, Schiffers remained part of this supportive network for all his life. Like many of his colleagues, he continued his research on Africa after 1945. With regard to the political relief strategies of German geographers, his concept of "relearning" and his imaginative geographies prove to be particularly interesting, especially since he continued his travels to Africa after 1945. Furthermore, he contributed his knowledge and his scientific skills to the Munich Institute and, until his death, was responsible for a variety of works on North Africa, including the previously mentioned work on Libya.

Schiffers' representation of Africa

While Schiffers' work can be divided into geographic and regional studies on the one hand and historical and political studies on the other, his entire work is characterized by continuity in his educational objectives, which is reflected in the places of publication, his narrative style, and last but not least his use of the media. There is also continuity at the core of his thinking, characterized by a political geography, as represented by the German geographer Friedrich Ratzel (1844–1904), which connects spatial arrangements and locations. Schiffers was interested in the interconnection of geographical conditions, the situation (*Lage*) according to Ratzel, life, survival, the "law of self-preservation", as he worded it (Schiffers-Davringhausen 1936, Forward). In his depiction, Africans – with clear racial differentiation – were and are as involved as Europeans in this fight against the geographical conditions in Africa. And for Schiffers, it was the Germans, especially German men, who had earned special merit in this fight in and for the region. Precisely in this fight, he saw the task of Germany in Africa, which he demanded in his works in the inter-war period regardless of the current political situation of Germany. Schiffers saw modern European technology in the form of cars and aeroplanes as a "saving" element in this "fight", which permitted men and women to control the landscape and let them have at least a partial victory.

This core of his thinking – the fate of geographical location and condition – is also found in his post-war works. In the preface to his monograph about the Sahel zone, which broached the issue of the drought and humanitarian catastrophe at the beginning of the 1970s, he argued that:

> The Sahel cannot escape the fate of the southern edge, of being situated in the immediate vicinity of the largest desert on earth. This is all the more true as direct interlocking with tongues of full desert aimed southward appears to be the rule. No matter how many expert reports and highly detailed aid projects are worked out, they will not change the fact that: Ignoring the natural geographic situation of the Sahel will result in illusion. (1976, Prologue)

The location and the threat by "desert tongues" was emphasized by an illustrative diagrammatic map with the memorable message that one cannot outrun such a threatening situation (Figure 1).

In contrast to his early writings, in his post-war scholarship, the idea of a German or European political "task" in Africa vanished. While the achievements of researchers and explorers remained a central topic, Schiffers now

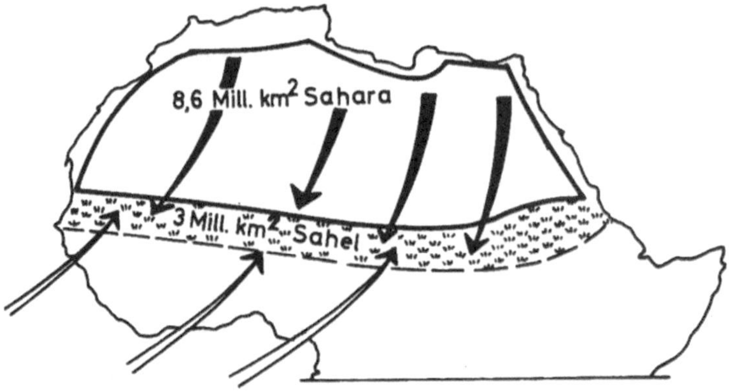

Vorspruch

Dem Südrand-Schicksal,
der Lage in der unmittelbaren Nachbarschaft
der größten Wüste der Erde,
kann der Sahel nicht entrinnen.
Dies umsomehr, als direkte Verzahnung
mit südgerichteten Vollwüsten-Zungen
die Regel ist.

Noch so viele fachwissenschaftliche Gutachten,
noch so diffizil ausgearbeitete Hilfsprojekte
ändern nichts an der Tatsache:

Illusionen macht sich,
wer am natürlichen Lageschicksal des Sahel
vorbeidenkt.

H. Sch.

Figure 1. The situation (Lage) of the Sahel presented as prologue by Schiffers in *Nach der Dürre. Die Zukunft des Sahel* (1976).

saw Europeans as being embedded in an emerging narrative about the past of European colonial policy. This shift becomes particularly apparent when comparing the two German editions of *The Quest for Africa*, the first of which was published in 1954, and an extended and updated edition published in 1962. Schiffers had already dealt with the current political situation of "revolts" and "underground movements" and the consequences of these for the whites in Africa in a final chapter of the first edition, linking his narrative with the exploration of the pre-European history of Africa – by Europeans (!). However, he envisioned a common future for Europeans and non-Europeans in Africa: "For the white man, and for the black, the African Adventure is by no means over" (1958, p. 345).

In the revised edition, which included the experience of the independence of African states and the decolonization process, Schiffers viewed the future and the past in a different light. He depicts the European colonial era as a stage in African history – certainly not an insignificant one, but preceded by a past of its own. And exploring this past was the responsibility of African scientists whom he saw leading the continent between tradition and the modern age into the future:

> Time and again, Europeans are amazed at the waves of cheers at independence festivities, the frenzy of the music and dances. Even though jet aircraft and shining Cadillacs, parades and a flood of conferences in ultra-modern structures herald breathless advancement – the voice of the drums is still alive. It reverberates in our ears. The pulse of the old continent is getting fresher and louder. African professors are looking into it and trying to find out what this heralds. How soon will we dig out an image of the unwritten black past, painted in graphic characters and worked out with the known source-critical and other new methods! The gnawing question what it was really like will then have found an *African* answer. (1962c, p. v)

Schiffers considered the exploration of the pre-European period as a crucial area for research, but for him the political challenge lay in managing the present and in the question of Africa's future. For the Africans were demanding answers to pressing questions "from a white guy who was changed himself by the flow of time". As Schiffers saw it, "Europeans like to act in accordance with formulas, as a walk through history proves. But the same (white) methods do not result in the same (black) outcomes everywhere" (1962c, p. 388).

The change Schiffers mentioned refers to the political transitions in the course of the African independence movements and in particular the decolonization process after World War II. It may not have meant active "relearning", but it did mean rethinking for Schiffers, because the time frame of his map of Africa widened. It now reached into the African past and extended to an uncertain future. On the other hand, he remained true to his nationalist and colonial thinking, because Schiffers held that the European "missions" had just interrupted the continent's own development at various spatial focuses, without playing down the violence and brutality of the European approach. Although he relied completely on the self-healing forces of the region or the African people and their actions in the post-war period, he could not solve a key problem in this perspective, which constitutes an area of tension, runs like a thread through his works, and remains linked with his narrative and geographical imagination. It is his enthusiasm for technology, since the use of instruments and means of transport such as the "motor vehicle made it possible to wipe away 2 million square kilometers of blank spots from the map of the black continent, which still in 1929 were next to the Nile basin that had been known for millennia" (1962c, p. 381). This technology had been brought to Africa by the Europeans, however, and therefore, despite all change brought about by the "white man", a dichotomy remained between African tradition and the onset of modern technology and forms of communication. In other words, despite his best efforts, he ultimately still imagined the African future as modernization along European lines.

In his study of Libya, Schiffers argued that, although the desert location was blessed with oil thanks to "Allah's providence", only modern oil extraction technology had removed the "barrier" function of the Libyan desert, had removed a "blind spot of the colonial era", launched the "project era" in Inner Africa, and facilitated the "bold Libyan reform experiment" (1975, pp. 10–11). Since Schiffers remained loyal to his specific geographical interests in Africa after 1945, the question arises as to how far his research on Libya was a particular challenge for a process of "relearning". One could even argue that the political system of an authoritarian military government under al-Gaddafi had a correspondence in his own political experiences with National Socialism. Nevertheless, the framework of his imaginative geography expanded and incorporated elements of an African potential for modernization, which in his earlier works had been exclusively based on European colonial powers. Even if Schiffers' imaginative geography remained partly in a tropical geography, a shift to a development-geography, referring to social, political and economic development, can be observed in the case of Libya.[12]

This shift in Schiffers' work towards a "development geography" as part of his process of "relearning", still coupled with the awe of a "German Kara Ben Nemsi style romanticism"[13] as Stowasser (1977, p. 220) called it, is also reflected in media use of pictures and maps. There was agreement among all reviewers that Schiffers' works featured a good picture selection and "instructive" maps (Plischke 1956, p. 371; Stowasser 1977, p. 219). The picture media, which in the first German edition of *Quest for Africa* consist of contemporary depictions, drawings, and photos, typically presented Europeans and European achievements in Africa – except for a few rock paintings – and support the narrative of a premodern, traditional Africa, especially because they are often found placed without comment inside the chapters.

The change between editions is interesting with respect to media use as well. The first German edition of the *Quest for Africa* was still populated by the "fantastic misshapen creatures from the dark continent", little-changed since Sebastian Münster had presented them to readers in his cosmography of 1550 (Figure 2). Schiffers, who was responsible for the compilation of the figures, uses this form of intertextual reference and to link to known images and ideas in order to underline the pre-modern character of Africa and also to emphasize the German title of his book (1954, p. 153; Münster 1545, p. 809). For example, the first page of the first edition (1954, p. 15) shows a "*kirangosi*", a group leader, in African headdress, who Burton had

Vielgestaltige Erkundungszüge in Oberguinea 153

Alles dieses macht es verständlich, daß die Erforschung aus einem vielgestaltigen Neben- und Nacheinander von Erkundungszügen bestehen mußte, an der Missionare wie Sklavenhändler, Agenten von privaten Handelsgesellschaften wie staatliche, außerdem militärische und friedliche Erkundungs-Expeditionen teilhatten.

Ihr erstes Kapitel endete höchst romantisch – für harmlose Gemüter wenigstens. Wir erleben die Herausbildung eines Eldorados für Seeräuber und andere Glücksritter aus aller Herren Ländern.

Phantastische Mißgestalten aus dem dunklen Erdteil
(Aus Seb. Münsters „Kosmographie", 1550)

Figure 2. "Fantastic misshapen creatures from the dark continent" taken from the Cosmography of Sebastian Münster (1550). From *Wilder Erdteil Afrika. Das Abenteuer der grossen Forschungsreisen* (Schiffers 1954, p. 153).

described and whose portrait can be found in numerous popular media of the nineteenth century, such as the popular journal *Gartenlaube* of 1890 (Figure 3). Eight years later, Schiffers (1962c, p. 3) replaced this portrait with the image of a terracotta head from the fifteenth century, an archaeological finding that points to Africa's handicraft, tradition, and its own past, rather than to a tropical present (Figure 4).[14]

Schiffers' ideas and his own "relearning" are also reflected in the photographic figures. The contrast becomes particularly evident when comparing the pre-war book *Silent Front*, published in 1936, to the second German edition of the *Quest for Africa* from 1962. The former underlines modernization by technology using photographic compositions. For example, the extension work on the "capital of the fascist colony of Libya" is illustrated by a view along an avenue lined with trees and leading to the Mediterranean, accompanied on the left by the old Ottoman fort and on the right by a modern high-rise building. No coincidence either is the photo of a young woman taken in front of a white bus in the desert, with the following caption: "Nothing documents the advance of European civilization more vividly than the young lady in desert attire" (Schiffers-Davringhausen 1936, p. 17).

Figure 3. "Kirangosi", the group leader as presented by Henry Morton Stanley. From *Wilder Erdteil Afrika. Das Abenteuer der grossen Forschungsreisen* (Schiffers 1954, p. 15).

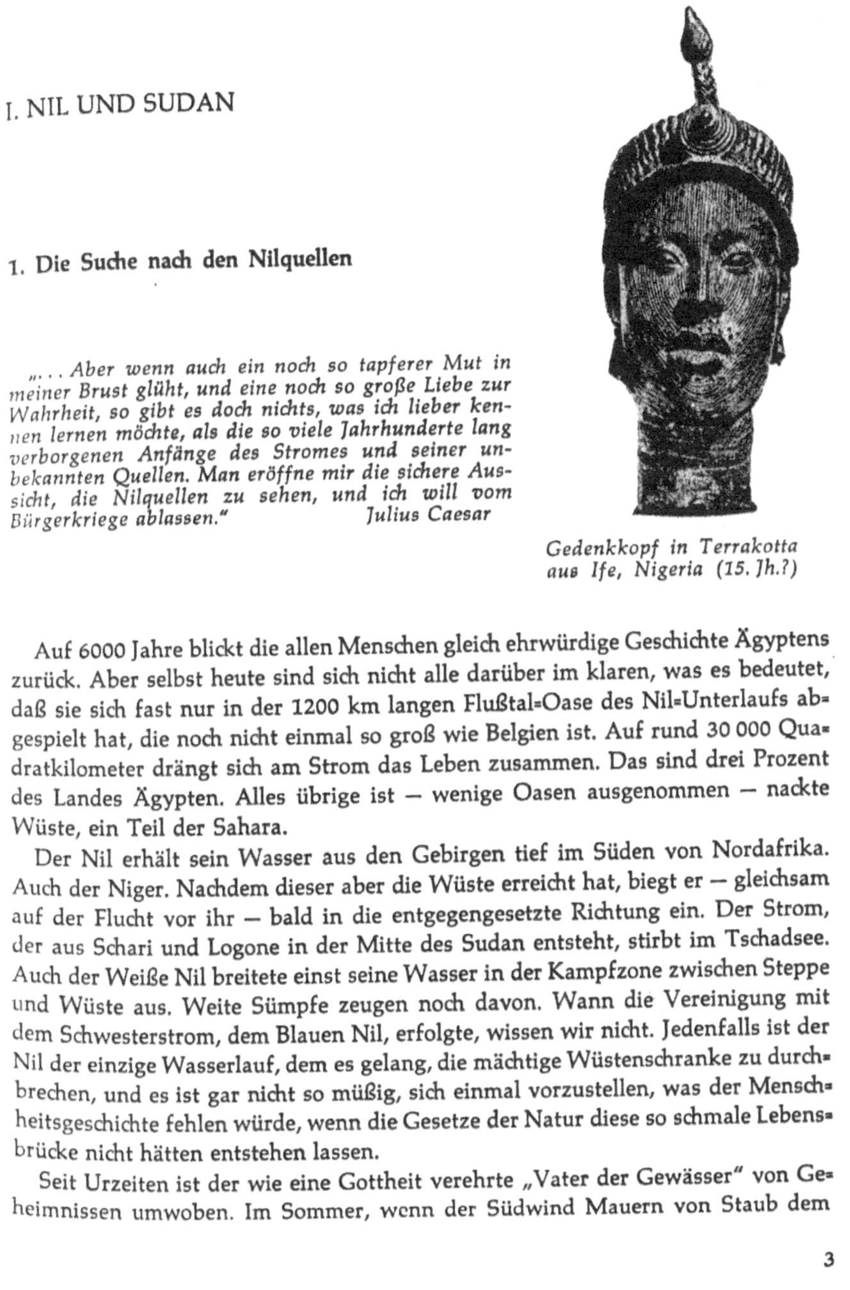

I. NIL UND SUDAN

1. Die Suche nach den Nilquellen

„. . . Aber wenn auch ein noch so tapferer Mut in meiner Brust glüht, und eine noch so große Liebe zur Wahrheit, so gibt es doch nichts, was ich lieber kennen lernen möchte, als die so viele Jahrhunderte lang verborgenen Anfänge des Stromes und seiner unbekannten Quellen. Man eröffne mir die sichere Aussicht, die Nilquellen zu sehen, und ich will vom Bürgerkriege ablassen." Julius Caesar

Gedenkkopf in Terrakotta aus Ife, Nigeria (15. Jh.?)

Auf 6000 Jahre blickt die allen Menschen gleich ehrwürdige Geschichte Ägyptens zurück. Aber selbst heute sind sich nicht alle darüber im klaren, was es bedeutet, daß sie sich fast nur in der 1200 km langen Flußtal-Oase des Nil-Unterlaufs abgespielt hat, die noch nicht einmal so groß wie Belgien ist. Auf rund 30 000 Quadratkilometer drängt sich am Strom das Leben zusammen. Das sind drei Prozent des Landes Ägypten. Alles übrige ist — wenige Oasen ausgenommen — nackte Wüste, ein Teil der Sahara.

Der Nil erhält sein Wasser aus den Gebirgen tief im Süden von Nordafrika. Auch der Niger. Nachdem dieser aber die Wüste erreicht hat, biegt er — gleichsam auf der Flucht vor ihr — bald in die entgegengesetzte Richtung ein. Der Strom, der aus Schari und Logone in der Mitte des Sudan entsteht, stirbt im Tschadsee. Auch der Weiße Nil breitete einst seine Wasser in der Kampfzone zwischen Steppe und Wüste aus. Weite Sümpfe zeugen noch davon. Wann die Vereinigung mit dem Schwesterstrom, dem Blauen Nil, erfolgte, wissen wir nicht. Jedenfalls ist der Nil der einzige Wasserlauf, dem es gelang, die mächtige Wüstenschranke zu durchbrechen, und es ist gar nicht so müßig, sich einmal vorzustellen, was der Menschheitsgeschichte fehlen würde, wenn die Gesetze der Natur diese so schmale Lebensbrücke nicht hätten entstehen lassen.

Seit Urzeiten ist der wie eine Gottheit verehrte „Vater der Gewässer" von Geheimnissen umwoben. Im Sommer, wenn der Südwind Mauern von Staub dem

3

Figure 4. Head made of terracotta from Ife, Nigeria, (15th century ?). From *Wilder Erdteil Afrika. Das Abenteuer der grossen Forschungsreisen* (Schiffers 1962c, p. 3).

All these pictures underline his message of Europeans bringing civilization to the savages. In case of the last mentioned one, the message was even gendered, since in the process of colonial conquest the women followed the men.

In the *Quest for Africa* of 1962, the photos and drawings presented by the author have a different function. In addition to the function of documenting, for example, archaeological findings (Figure 5), and in this way giving an image to the past, they point out the continuity of tradition in social life and the change and modernization of Africa, for example, with respect to urbanization, either by their comparative arrangement or by respective captions (Figure 6) (1962c, figs 25–28, pp. 65–71).

Schiffers uses this technique much more frequently in his book on Libya of 1975. Photographs similarly document the past and present and put traditional lifestyles next to modern urbanity and industrialization. He makes a clear distinction between traditional forms of social life and traditions in business life, which hardly ever are viewed as idyllic but clearly as obsolete, "belonging in a museum" (1975, pp. 118, 155). Even though there are occasional remarks critical of modernism (for example, when he points out the "monotony" of modern housing settlements), his dichotomic view of Libya is reflected in a new layout feature in this work. For the first time, Schiffers uses color photographs and, except for an Arabic cemetery, these photos exclusively show the developed "new south in the desert" with its modern cities, the "new face of the desert" with irrigated green areas, the new university, commercial ports and the burning oil fields in the desert (also shown on the cover) (1975, pp. 113, 120). The color underlines contrast; a green desert is relatively hard to see in a black and white photograph, and the color photographs bring out the natural resources in the desert soil and the modernization of the country.

Far beyond the pictorial material outlined here, all of Schiffers' publications are also characterized by extensive maps. These maps differ in various respects from those in the *Atlas of Africa* Priority Program published at the same time (1975, pp. 118, 155). Since Schiffers had drawn and selected the maps for *Silent Front* and for the two German editions of the *Quest for Africa* himself, the question arises to what extent his "relearning" is also reflected in maps. In general, all his maps are in black and white. Most of them are sections based on manual drawings or featuring handwritten entries.

An overview of (almost) the entire continent is provided on the covers of both of the German editions of *Quest for Africa*, with significant differences. In the first edition (1954), Schiffers showed the continent in two parts with a cutting line on the level of Zanzibar, whereas the Mediterranean and the countries surrounding it are shown in relation to Africa and the map ends in the east with today's Iran (Figure 7). In the revised edition, there is just one map of the same scale showing the entire continent but no surrounding countries. Completely isolated, only surrounded by the sea, with no connections to Europe or Asia, the continent and its multitude of countries stands for itself. There is only one map on the front cover (Figure 8).This island map on the book cover shows, in hatching, the former German dependencies and the

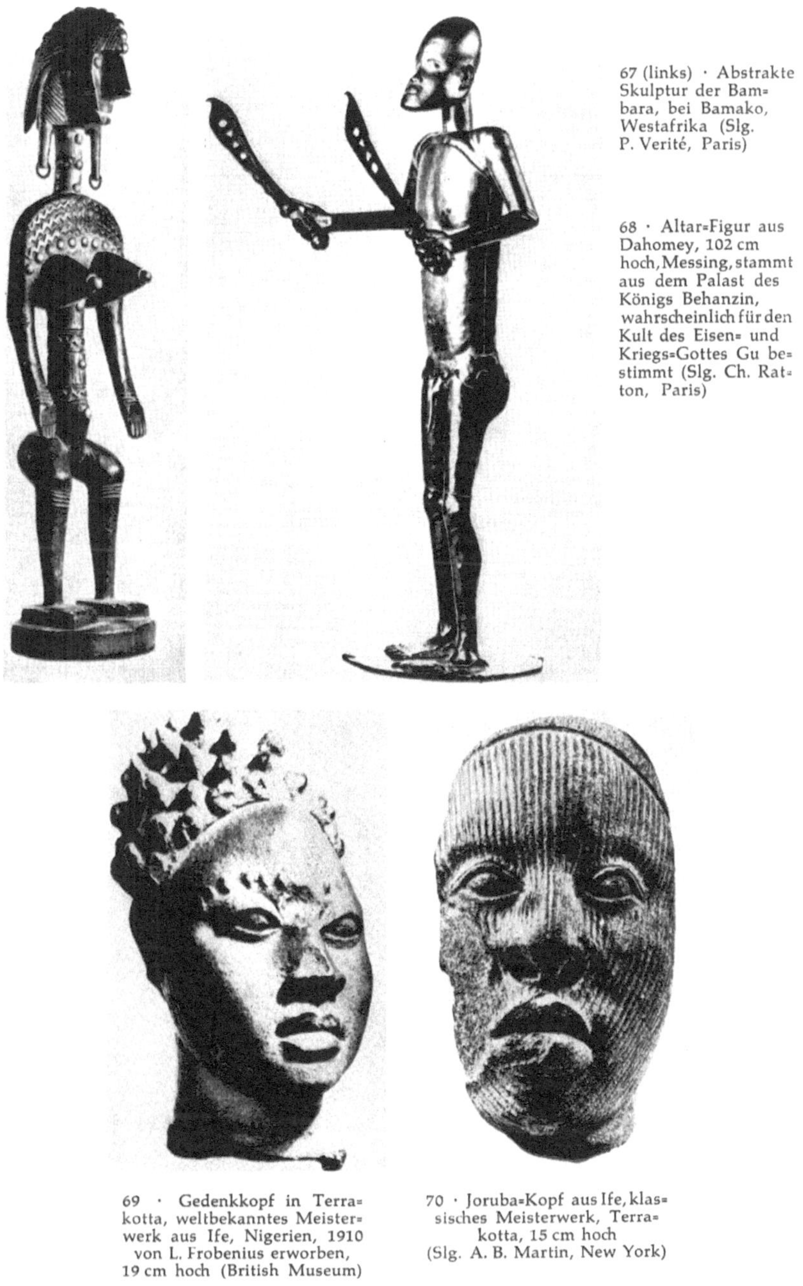

Figure 5. Pre-European African sculptures in private collections and museums in Europe and US. From *Wilder Erdteil Afrika. Das Abenteuer der grossen Forschungsreisen* (Schiffers 1962c, figs 67–70).

27 · Die Stadt Kumasi an der Goldküste 1870

28 · Kumasi 1958

Figure 6. City of Kumasi, Ghana, in 1870 and 1958. From *Wilder Erdteil Afrika. Das Abenteuer der grossen Forschungsreisen* (Schiffers 1962c, figs 27–28).

territories, which were not under British or French rule (!). However, the numerous white spots of the continent, in turn, "veil" the colonial past when they are presented to the reader, without further explanation, with their current political independence, as in the case of Niger, Kenya or South Africa.

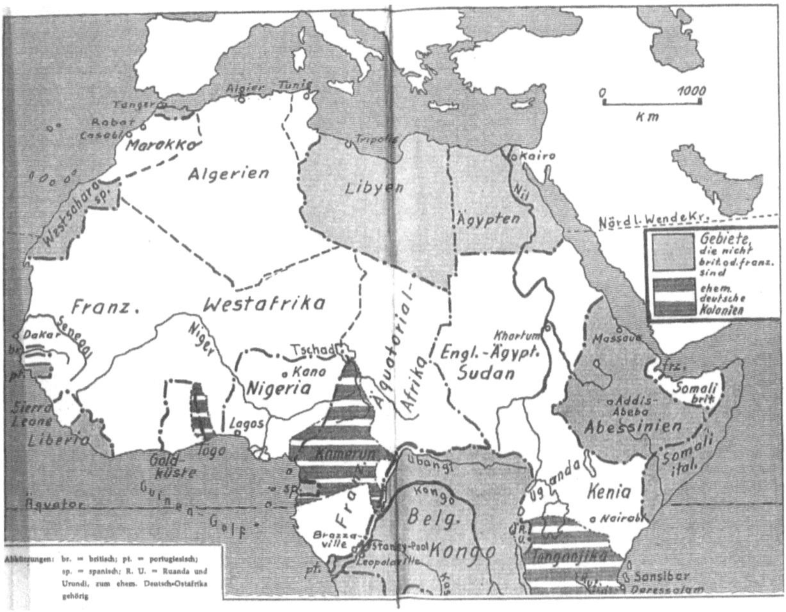

Figure 7. Africa, Europe and the Arabian Peninsula presented on the front cover of *Wilder Erdteil Afrika. Das Abenteuer der grossen Forschungsreisen* (Schiffers 1954).

The colonial past was also highlighted eight years before. However, the interesting thing about this divided map on the book cover is not only the connection to, and the perspective towards, Europe, but also the different time layers. Whilst the two maps on the cover illustrate the colonial rule in a mix of the past and present, another map in the text refers to the process of "unveiling" with two sections in time taken from 1850 and 1890. Here, the white spots indicate the unexplored territories of the African continent, referring with the white color to an established and, since the late nineteenth century, quite popular understanding. The mixture of the same white spots for completely different historical and political references within the same book confront the reader with a puzzle picture and defamiliarizes the "popular" view of Africa (see in contrast the map on "unveiling" in 1954, p. 63).

Two more features of the numerous maps in both works and the *Silent Front* are notable. Almost all produce strikingly dynamic images through the depiction of the movements of explorers, as well as military expansions such as the Italian advance. They illustrate multiple time layers at once, for example, when the journey by Friedrich Gerhard Rohlfs (1831–1896) is shown in the same map as the "conquest of Libya" by the Italians in 1931 (1936, p. 272). The advance of the Italians marked in "bold and black", in addition shown in Sütterlin script, and the refugee movements it caused produce a dynamic effect in the map image, which is otherwise often a

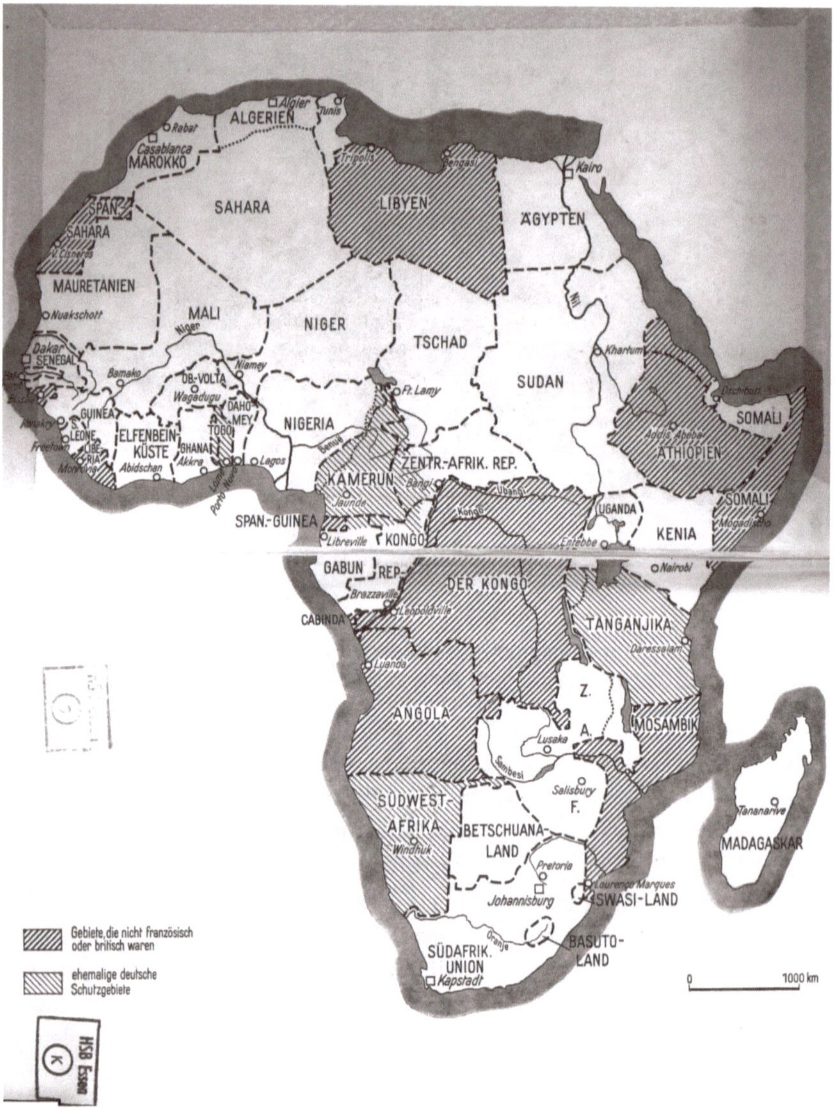

Figure 8. The African continent presented on the front cover of *Wilder Erdteil Afrika. Das Abenteuer der grossen Forschungsreisen* (Schiffers 1962c).

comparatively static medium (Figure 9). While Schiffers substantially focused on the European past, the change in perspective outlined above towards the pre-European time, the mixing of time layers, and the dynamic, are also found in the second German edition of the *Quest for Africa*. In a detail map without a frame titled "African civilizations of pre-European time in the east and south east of the continent", he marked spent resources, meteorological phenomena, trade routes, and products in addition to topographical

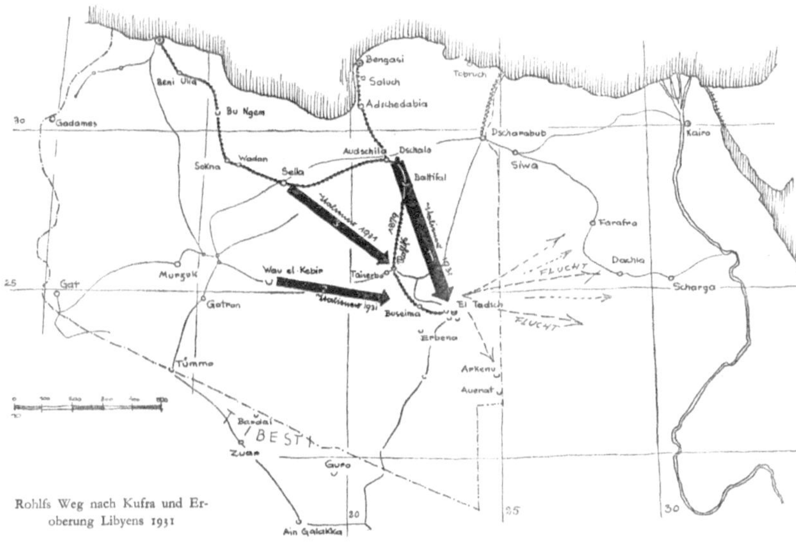

Figure 9. "Rohlfs way to Kufra [dotted line] and conquest of Libya in 1931[black arrows]" presented in *Stumme Front. Männer und Mächte im Banne der Sahara* (Schiffers-Davringhausen 1936, p. 272).

features, but these are missing any temporal classifications (1962c, p. 111) (Figure 10).

The map images are fundamentally different from the scientific mapping in the *Atlas of Africa* Priority Program, but in a certain way they correspond to the cartographic techniques and forms of visualizations of these years. The fact that Schiffers was familiar with the techniques and map languages of his time is above all reflected in this dynamic approach, which was a popular cartographic design of the inter-war period for organizing spaces (Haslinger and Oswalt 2012; Jureit 2012; Tilley 2011). This familiarity with the techniques also becomes apparent when looking into the Libya book, which is compared to these earlier maps predominantly characterized by a "more scientific" map language, although some "drafts" made by Schiffers can still be found there as well (1975, p. 165). Different time domains, such as "classical antiquity" and "modern era" in one map are the exception; instead, the future is depicted more positively, with planning and project maps (1975, pp. 48, 149). But even more than that Schiffer's maps are cast into a model that is based on Walter Christaller's space theories and visualizations of central locations, which attained great popularity in the planning contexts after World War II (Christaller 1968; Barnes and Minca 2013; Trezib 2014).

In addition – and this is a special feature – Schiffers has included an Arabic planning map in his work, which is indicated by its characters and caption. This is not astonishing at first glance, considering the change in narrative

16. Am Kreuzweg der Jahrtausende

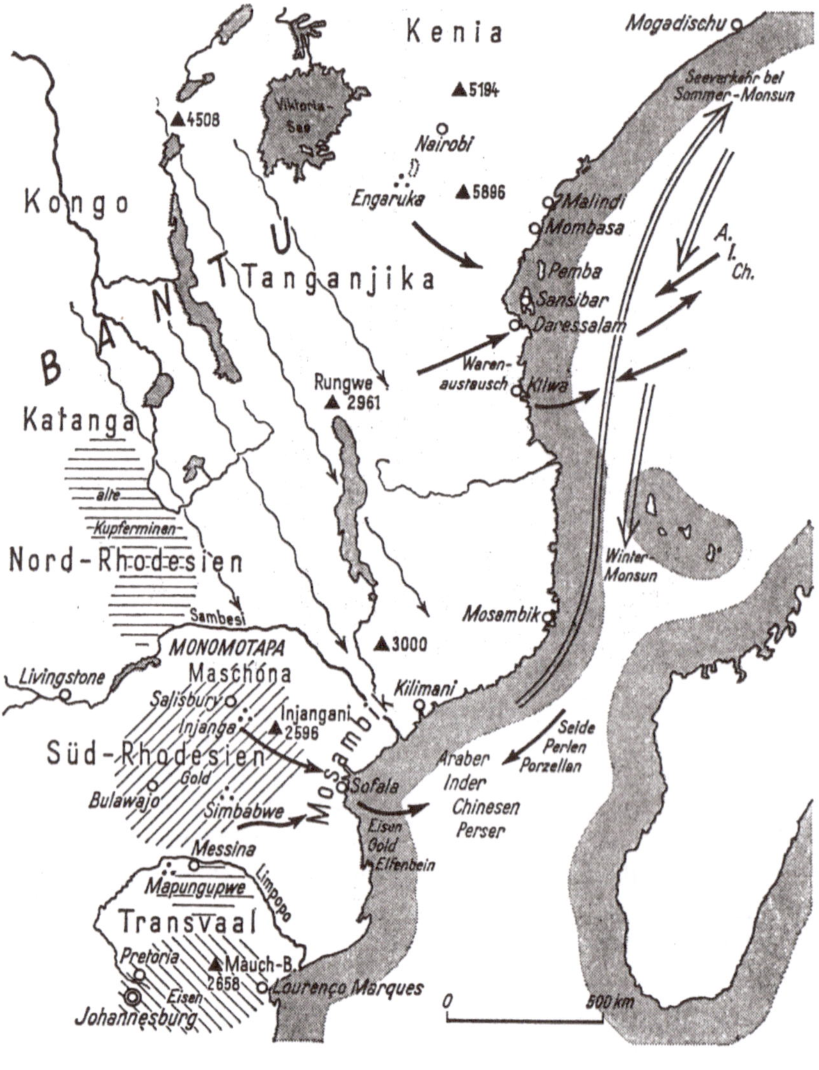

Karte 7 · Afrikanische Kulturen der Vor=Europäer=Zeit im Osten und Südosten des Kontinents

Das weite Land zwischen dem Viktoria=See und Transvaal ist stellenweise, bis auf die vorgelagerten Inseln hinaus, förmlich übersät von bemerkenswerten Zeugnissen vergangener Kulturen: Reste von Dörfern und Stadtsiedlungen, von Palästen und Gräbern von Brunnen und Straßen, von kunstvollen Befestigungen und unglaublich weit gedehnten Terrassenanlagen, von Bergwerksanlagen auf Gold, Kupfer und Eisen.

Figure 10. African Cultures in Pre-European Times, in *Wilder Erdteil Afrika. Das Abenteuer der grossen Forschungsreisen* (Schiffers 1962c, p. 111).

presented here, which is consistently reflected in the text, figures, and maps, since his focus is on Arab-African potentials. But in among many of his colleagues, who had booted him out at the DFG and kicked him out of the *Atlas of Africa* Priority Program, this was not a matter of course. In the mid-1970s, it was not common among geographers to disclose African knowledge, let alone give it such a prominent place.

"Relearning" of an imaginative geography?

For Schiffers, the term "relearning" was associated with a didactic impetus, characterizing not only his own experience after 1945 but also addressed against the Eurocentrism of his readership. He had self-awareness, which he explicitly formulated, namely that his views of Africa had also undergone a change, which was not without consequences for his imaginative geography of Africa. The remapping of his imaginative geography took place in two ways. Firstly, his map extended the temporal dimension of African history prior to European exploration, and secondly his acceptance of African actors who had their own lifestyles, practices, and interests which differed from European ideas and expectations.

"Relearning" thus meant for Schiffers the departure from a colonial African geography, as he had portrayed since the inter-war period, and the endeavor to accept African history and development potential in the present and future. In doing so, he linked to international debates. On the other hand, with his conception of a dichotomy of tradition and modernity and a necessary modernization according to the Western model, his text and images still evoked the geopolitics of a colonial geography (see Jones 1987; Fauvelle-Aymar 2013). Heinrich Schiffers was not alone in his remapping of his imaginative geography. Political developments in Africa and the international debates and programs for development aid also led some British and French geographers such as Buchanan, McGee, and to some extent even Yves Lacoste to relearn their imaginative geographies.

In this sense, Schiffers' path toward a "political" development geography was not unique, but there were other ways of "relearning" in Germany, as evidenced by a contemporary, who gained great media attention in the Federal Republic and who also contributed substantially to West German knowledge about Africa. The Frankfurt Zoo director, Bernhard Grzimek[15] (1909–1987), who had been bringing African wildlife into German living rooms since the late 1950s with documentaries and spectacular television appearances, also remapped Africa (Torma 2004). His understanding of African nature as a peaceful encounter between humans and animals broke with the colonial practice of big game hunting that demonstrated white power over men, animals and territories (Gißibl and Paulmann 2013, pp. 103–105). Common to Schiffers and Grzimek was a geopolitical core of thinking which sought to explain circumstances and developments from the conditions of space and situation. However, the "relearning" of the two resulted in virtually contrasting imaginative geographies which had their origins in the colonial and spatial debates of the 1920s and 1930s, and which can be broadly outlined by the antagonisms of nature versus culture and tradition vs. the modern age. Bernhard Grzimek therefore filled his revised mental map with the incredible nature and wildlife he saw endangered not only by Europeans, but equally by the African people. In his opinion, the threat arose from progress, particularly of American provenance, with the spatial expansion of man and his consumption at the expense of nature (Torma 2004, pp. 42–54). For Schiffers, who firmly insisted on not conveying any "fashionable geological or ecological clichés" (1975, p. 11) in his works, the potential for action and for the future of the continent lay precisely in the spread of the modern age with its technical advances based on African resources such as, for instance, coal and petroleum, in order to overcome adverse natural conditions. Whereas Grzimek therefore underlined his criticism of progress by highlighting wild, unspoiled nature, Schiffers outlined the issues of resources, which were already associated with the colonial policy, replacing European actors with African actors in his mental map, while still maintaining his technological focus. In terms of map images, Grzimek's map resembled the maps of Africa which were to be found in numerous children's atlases until well into our time and which reduced Africa to jungles and wildlife, whilst Schiffers' map shows towns and cities, industry and oil fields, exploited by African people. Against the backdrop of this comparison, "relearning" turns out not to be a complete remapping of Schiffers' imaginative geography, but rather a displacement and enrichment of existing elements.

Notes

1. The term *Fourth World* was coined in the late 1960s and used by Schiffers and others to differentiate between poor and even poorer countries.
2. Heinrich Schiffers' curriculum vitae of 25 July 1944, in University Archives Marburg, 307d No. 883.
3. Bundesarchiv Bestand NS 15, Der Beauftragte des Führers für die Überwachung der gesamten geistigen und weltanschaulichen Schulung und Erziehung der NSDAP. Landesarchiv Düsseldorf NW 1048-33.
4. Employees under Kanter's professorship also included the geographer Kurt Scharlau (1906–1964), a member of the Waffen SS (the military part of the SS), who worked in Marburg again after 1951. For information about both, see Klee (2003, pp. 298, 527), Kanter 1967, see Warren's (1968) review of this book.
5. For the history of the Athenäum publishing house and the publisher's involvement in National Socialism, see Körner (2002, 2003),
6. Bundesarchiv Koblenz, B 227 Afrikakartenwerk, Signature 212285, Volume 1, Letter by Paul Junker of 30 October 1961.

7. Walther Manshard (b. 1923) had qualified as a professor in 1959 under the geographer Kurt Kayser (1905–1984), who subsequently assumed overall control of the DFG's *Atlas of Africa* Priority Program (Schwerpunktprogramm Afrika-Kartenwerk), and was appointed to the Gießen chair in 1963. Manshard was also involved in the *Atlas of Africa* from the outset and, in later years, worked for both UNESCO and the United Nations (Manshard 1998).
8. Bundesarchiv Koblenz, B 227 Afrikakartenwerk, Signature 212285, Volume 1, Letter by Paul Junker of 30 October 1961.
9. Bundesarchiv Koblenz, B 227 Afrikakartenwerk, Signature 212285, Volume 1, Letter by Paul Junker of 30 October 1961.
10. Bundesarchiv Koblenz, B 227 Afrikakartenwerk, Signature 212285, Volume 1, Letter by Paul Junker of 30 October 1961.
11. Landesarchiv Düsseldorf NW 1048-33.
12. A similar process has been described for the British geographer Buchanan, who also became a supporter of development policy in the late 1960s. (Power and Sidaway 2004) The remarkable parallels, not least with regard to the position at the margin of the discipline and the political influence, still require further comparative investigations.
13. Kara Ben Nemsi is the first-person narrator and hero of several novels of the German author Karl May (1842–1912). In his adventure novels, the young German Kara Ben Nemsi travelled with a few companions through North Africa, Middle East, Sudan, and the Balkans. The novels have had wide circulation in Germany and popularized an imaginative geography of the Orient. Modern research on May emphasizes that he propagated ideas of European superiority as he informed his readers about areas including about the Ottoman Empire.
14. It would be as instructive to compare the representations of animals, which have to remain unconsidered here.
15. I would like to thank Matthias Middell, Leipzig, for the reference and comparison with Bernhard Grzimek.

Acknowledgements

I would like to thank Norbert Götz, Janne Holmén, Steven M. Schnell and the anonymous referees for their helpful suggestions on different drafts of this paper. Ramona Gläser supported me with her always careful and reliable research.

Disclosure statement

No potential conflict of interest was reported by the author.

References

Aly, G. and Heim, S., 1993. *Vordenker der Vernichtung. Auschwitz und die deutschen Pläne für eine neue europäische Ordnung.* Frankfurt am Main: Fischer-Taschenbuch-Verl.

Andreasson, S., 2005. Orientalism and African development studies: the "reductive repetition" motif in theories of African underdevelopment. *Third World Quarterly*, 26 (6), 971–986.

Ansprenger, F., 1967. African studies in the Federal Republic of Germany. *The Journal of Modern African Studies*, 5, 401–406.

Barnes, T. and Minca, C., 2013. Nazi spatial theory. The dark geographies of Carl Schmitt and Walter Christaller. *Annals of the Association of American Geographers*, 103, 669–687.

Böhm, H., 2003. Annäherungen. Carl Troll (1899–1975) – Wissenschaftler in der NS-Zeit. *In*: M. Winiger, ed. *Carl Troll: Zeitumstände und Forschungsperspektiven.* Sankt Augustin: Asgard-Verl., 1–99.

Büschel, H., 2014. *Hilfe zur Selbsthilfe. Deutsche Entwicklungsarbeit in Afrika 1960–1975.* Frankfurt am Main: Campus-Verl.

Butzer, K.W., 2004. Practicing geography in a totalitarian state: (re)casting Carl Troll as a Nazi collaborator? *Die Erde*, 135, 223–231.

Christaller, W., 1968. *Die zentralen Orte in Süddeutschland. Eine ökonomisch-geographische Untersuchung über die Gesetzmässigkeit der Verbreitung und Entwicklung der Siedlungen mit städtischen Funktionen.* 3rd ed. Darmstadt: Wiss. Buchges.

Cosgrove, D.E., 2008. *Geography and vision. Seeing, imagining and representing the world.* London: I.B. Tauris.

Eckert, A., 2015. Westdeutsche Entwicklungszusammenarbeit mit Afrika. Ein Blick auf die 1950er bis 1970er Jahre. *Deutsche Zeitgeschichte transnational.* Göttingen: Wallstein Verlag, 27–44.

Fauvelle-Aymar, F.-X., 2013. *Le rhinocéros d'or. Histoires du Moyen Âge africain.* Paris: Alma éditeur.

Gieseking, J.J., 2013. Where we go from here. *Qualitative Inquiry*, 19 (9), 712–724.

Gißibl, B. and Paulmann, J., 2013. Serengeti darf nicht sterben. *In*: J. Zimmerer, ed. *Kein Platz an der Sonne. Erinnerungsorte der deutschen Kolonialgeschichte.* Frankfurt am Main: Campus-Verl., 96–108.

Gregory, D., 1994. *Geographical imaginations.* Cambridge: Blackwell.

Haar, I. and Fahlbusch, M., 2005. *German scholars and ethnic cleansing, 1919–1945.* New York: Berghahn Books.

Haslinger, P. and Oswalt, V., eds., 2012. *Kampf der Karten. Propaganda- und Geschichtskarten als politische Instrumente und Identitätstexte.* Marburg: Verl. Herder-Inst.

Hellen, J.A., 1968. Review of Heinrich Barth. Ein Forscher in Afrika. Leben, Werk, Leistung. *The Geographical Journal*, 134 (1), 105.

Hellen, J.A., 1970. Review of Gustav Nachtigal 1869–1969. By Theodor Heuss, Herbert Ganslmayr and Heinrich Schiffers, Berlin 1969. *The Geographical Journal*, 136 (1), 148.

Hellen, J.A., 1974. Review of Die Sahara und ihre Randgebiete. Darstellung eines Naturgrossraumes. Edited by Heinrich Schiffers, München 1971–1973. *The Geographical Journal*, 140 (2), 312–313.

Jacob, E.G. ed., 1938. *Deutsche Kolonialpolitik in Dokumenten. Gedanken und Gestalten aus den letzten 50 Jahren.* Leipzig: Dietrich.

Jacob, E.G. 1956. Review of Heinrich Schiffers. Wilder Erdteil Afrika. *Historische Zeitschrift*, 181, 232–233.
Jones, A., 1987. Raw, medium, well done. A critical review of editorial and quasi-editorial work on pre-1885 European sources for sub-Saharan Africa, 1960–1986. Madison, WI: African Studies Program, University of Wisconsin – Madison.
Jureit, U., 2012. *Das Ordnen von Räumen. Territorium und Lebensraum im 19. und 20. Jahrhundert*. Hamburg: Hamburger Ed.
Kanter, H., 1967. Libyen / Libya Eine geographisch-medizinische Landeskunde / A geomedical monograph. *Medizinische Länderkunde / Geomedical Monograph Series, Beiträge zur geographischen Medizin/Regional Studies in Geographical Medicine 1*. Berlin: Springer.
Kitchin, R. and Blades, M., 2001. *The cognition of geographic space*. London: I. B. Tauris.
Kitchin, R. and Freundschuh, S., eds., 2000. *Cognitive mapping. Past, present and future*. London: Routledge.
Klee, E., 2003. *Das Personenlexikon zum Dritten Reich. Wer war was vor und nach 1945?* Frankfurt: S. Fischer.
Körner, K., 2002. "Verlorene Siege" (I): Der Junker und Dünnhaupt Verlag 1927–1945. *Aus dem Antiquariat. Zeitschrift für Antiquare und Büchersammler*, 3 (26. März 2002), 130–141.
Körner, K., 2003. "Verlorene Siege" (II): Der Athenäum-Verlag von 1949–1989. *Aus dem Antiquariat. Zeitschrift für Antiquare und Büchersammler*, 2, 83–102.
Manshard, W., 1998. *Als Geograph auf internationaler Bühne. Eine autobiographische Skizze*. Saarbrücken: Verl. für Entwicklungspolitik.
Maxwell, A.J., 1959. The quest for Africa by Heinrich Schiffers. *World Affairs*, 122 (4), 128.
Münster, S., 1545. *Cosmographia. Beschreibung aller Lender. Cosmographia universalis*. Basel: Petri.
Plischke, H., 1956. Schiffers, H.: Wilder Erdteil Afrika (Book Review). *Orientalistische Literaturzeitung*, 51, 371.
Power, M. and Sidaway, J.D. 2004. The degeneration of tropical geography. *Annals of the Association of American Geographers*, 94 (3), 585–601.
Schenk, F.B., 2013. Mental maps. Die kognitive Kartierung des Kontinents als Forschungsgegenstand der europäischen Geschichte. *Europäische Geschichte Online (EGO) hg. vom Leibniz-Institut für Europäische Geschichte Mainz*. Available from: http://ieg-ego.eu/en/threads/theories-and-methods/mental-maps/frithjof-benjamin-schenk-mental-maps-the-cognitive-mapping-of-the-continent-as-an-object-of-research-of-european-history [Accessed 24 November 2017].
Schiffers, H., 1944. *Die Tiniri als Typus eines nordafrikanischen Wüstenraumes*. Thesis (PhD). Marburg University.
Schiffers, H., 1952. *Die grosse Reise: Dr. Heinrich Barths Forschungen und Abenteuer 1850–1855*. Minden (Westf.): Köhler.
Schiffers, H., 1953. *Mein Erdkundebuch: Mit vielen Kt. u. Abb.* Düsseldorf: Bagel.
Schiffers, H., 1954. *Wilder Erdteil Afrika. Das Abenteuer der grossen Forschungsreisen*. Bonn: Athenäum-Verl.
Schiffers, H., 1957. *The quest for Africa [Wilder Erdteil Afrika]*. Two thousand years of exploration. Translated by Diana Pyke. London: Odhams Press.
Schiffers, H., 1958. *The quest for Africa. Two thousand years of exploration*. 1st American ed. New York: Putnam.
Schiffers, H., 1962a. *Afrika. Harms Erdkunde in entwicklender anschaulicher Darstellung*. 6th/7th ed. München: Paul List.

Schiffers, H., 1962b. *Libyen und die Sahara*. Bonn: Schroeder.
Schiffers, H., 1962c. *Wilder Erdteil Afrika. Das Abenteuer der grossen Forschungsreisen*. 2nd rev. and enl. ed. Bonn: Athenäum-Verl.
Schiffers, H., 1975. *Libyen. Brennende Wüste – blühender Sand* [Unter Mitarb. von Redmer H. and Weis H.]. Berlin: Safari-Verl.
Schiffers, H., 1976. *Nach der Dürre. Die Zukunft des Sahel*. München: Weltforum-Verlag.
Schiffers-Davringhausen, H., 1935. *Menschen unter Allahs Sonne. Dt. Augen sehen Nordafrika*. Berlin: Wegweiser-V.
Schiffers-Davringhausen, H., 1936. *Stumme Front. Männer und Mächte im Banne der Sahara*. Leipzig: Goldmann.
Schiffers, H. and Wagner, J., 1973. *Harms Handbuch der Erdkunde: Harms Erdkunde in entwickelnder, anschaulicher Darstellung*, Vol. 5: Afrika. München [u.a.]: List.
Schiffers, H., Wagner, J., and Eggers, W., 1957. *Harms' Handbuch der Erdkunde in entwickelnder, anschaulicher Darstellung*, Vol. 4: Afrika. 5th ed. Frankfurt: Atlantik-Verl. List.
Schneider, U., 2012. Wartezeit beendet. Das "Afrika-Kartenwerk" der Deutschen Forschungsgemeinschaft. *In*: S. Günzel and L. Nowak, eds. *Kartenwissen. Territoriale Räume zwischen Bild und Diagramm*. Wiesbaden: Reichert, 246–264.
Schrafstetter, S., 2008. Verfolgung und Wiedergutmachung. Karl M. Hettlage: Mitarbeiter von Albert Speer und Staatssekretär im Bundesfinanzministerium. *Vierteljahrshefte für Zeitgeschichte*, 56 (3/2008), 431–466.
Schultz, H.-D., 1989. Versuch einer Historisierung der Geographie des "Dritten Reiches" am Beispiel der geographischen Großraumdenkens. *In*: P. Jüngst, K. Pfromm, and H.-J. Schulze-Göbel, eds. *Urbs et Regio. Geographie und Nationalsozialismus*. Kassel: GHK, 1–76.
Smith, T.R. and Black, L.D., 1946. German geography: war work and present status. *Geographical Review*, 36 (3), 398–408.
Stowasser, K., 1977. Schiffers "et al.": Libyen: Brennende Wüste – Blühender Sand (Book Review). *Middle East Journal*, 31 (2), 219.
Tilley, H., 2011. *Africa as a living laboratory. Empire, development, and the problem of scientific knowledge 1870–1950*. Chicago: University of Chicago Press.
Torma, F., 2004. *Eine Naturschutzkampagne in der Ära Adenauer. Bernhard Grzimeks Afrikafilme in den Medien der 50er Jahre*. München: Meidenbauer.
Trezib, J., 2014. Transnationale Wege der Raumplanung. Der israelische Nationalplan von 1951 und seine Rezeption der Theorie "zentraler Orte". *Zeithistorische Forschungen/Studies in Contemporary History*, 11, 11–35.
Troll, C. 1947. Die geographische Wissenschaft in Deutschland in den Jahren 1933 bis 1945. Eine Kritik und Rechtfertigung. *Erdkunde*, 1, 3–48.
Troll, C. 1949. Geographic science in Germany during the period 1933-1945: A critique and justification. *Annals of the Association of American Geographers*, 39 (2), 99–137.
Warren, A., 1968. Reviewed work(s): Libyen (Libya) by Helmuth Kanter; Libyen (Libya) Geomedicalmonograph Series No. 1 by H. J. Jusatz. *The Geographical Journal* 134, 104.
Werlen, B., 2008. *Sozialgeographie. Eine Einführung*. 3rd rev. and enl. ed. Bern: Haupt.
Zimmerer, J., 2004. Im Dienste des Imperiums. Die Geographen der Berliner Universität zwischen Kolonialwissenschaften und Ostforschung. *Jahrbuch für Universitätsgeschichte*, 7, 73–100.

Palestine, Israel, Gaza, and the West Bank: the muddled mental maps of French and Israeli students

Efrat Ben-Ze'ev and Chloé Yvroux

ABSTRACT

The Palestinian–Israeli Conflict is perceived by many – observers and parties to the conflict alike – as a struggle of two peoples over the same land. Yet, through this century-long conflict (and more so as Israel has expanded and deepened its occupation), what was once, perhaps, imagined as a single land has become an assortment of territories. These territories bear multiple names and different legal statuses, and their boundaries are often blurred. In light of the jumbled patchwork that Palestine–Israel has become, we examine the ways that the conflict's territorial dimensions are imagined and represented. We study the mental maps of the region held by higher education students from Israel, both Jewish and Arab-Palestinian, as well as with university students from Montpellier, France. The representations indicate that while the French students were almost completely at a loss regarding the conflict's spatial dimensions, the students from Israel were also confused, especially regarding the Occupied Palestinian Territories. We argue that these findings stem from a wider process of deterritorialization, linked to the conflicting relations between state and nation and intensified by a policy of chaotic spatial arrangements.

Introduction

The Palestinian–Israeli Conflict is perceived by many – observers and parties to the conflict alike – as a struggle of two peoples over the same land. Yet, through this century-long conflict (and more so as Israel has expanded and deepened its occupation), what was once, perhaps, imagined as a single land has become an assortment of territories. These territories bear multiple names and different legal statuses, and their boundaries are often blurred. In light of the jumbled patchwork that Palestine–Israel has become, we examine the ways that the conflict's territorial dimensions are imagined and represented. Much of our focus here is on the representation of the Occupied Palestinian Territories (OPT) of the West Bank.

In late 1947, the United Nations decided on the partition of Palestine into a Jewish and an Arab state. Yet, while the state of Israel is celebrating its 70th anniversary, a state for the Arabs of Palestine has yet to be established. Moreover, the OPT of the West Bank, with 2.7 million inhabitants under military rule, is continuously settled by Israeli-Jews with a wall, fences, and roadblocks fragmenting the territory in unparalleled ways. In contrast, Israel unilaterally retreated from the Occupied Gaza Strip, with 1.5 million inhabitants in 2005, yet maintains rule over its perimeter. With such a variety of control methods and territorial arrangements, how is the spatiality of this conflict perceived?

We asked French and Israeli students to sketch their mental maps of the "country" (in the case of Israelis) and this "territory" (in the case of the French), leaving it up to them to decide what it means, and then talked about their understanding. We opted for two different subject positions (Kitchin and Blades 2002). One was that of the French students, who viewed the conflict as "outsiders" and gathered their information through secondary sources such as the media, education, storytelling, and books; they had no first-hand experience of Palestine/Israel. They were outsiders but certainly not detached since the conflict is prominent in French daily lives. French public interest is evident in the frequent media coverage and the high number of solidarity and human rights organizations, publications, and controversies on the topic; some even call it a "passion française" (Sieffert 2004). In contrast, the students in Israel are part of the conflict and have an immediate spatial connection to it.

Despite the significant differences, both for the French and the Israelis, the understanding of this territory is a mediated one. In other words, their perceptions are molded through certain lenses. Such lenses operate on us, the authors, hence our own subject position is also embedded in our research approach as well as in the way we represent our findings. Being well aware of the divergent manners through which this conflict is represented, we hope to uncover the spatial perceptions that characterize two different parties – one that is part of it and another, a spectator. We ask what the collective images of the territory within each of the two groups are, and agree with Saarinen who wrote that "it seems important in a world continually upset by international conflicts, to try to gain an understanding of variations in world views" (1973, p. 148).

While we dwell on the variations, our discoveries also point to a commonality. The OPT, and particularly the West Bank, are an amorphous entity for both groups. The West Bank neither has a single name nor does it fit our participants' familiar spatial frameworks. It is often understood as lacking shape and clear substance, almost deterritorialized, a term that we will further elaborate on in the following section.

Mental maps, conflict, and deterritorialization

From its early phases, the study of mental maps aspired to reach applicable conclusions that would improve people's lived experience.[1] Lynch (1960), working in American cities in the post-World War II period, asked how people could decipher the urban spaces they inhabit in order to feel more at ease.[2] Slightly later, Guy Debord (in Knabb 1981) was mapping the ambience of Parisian neighborhoods with the hope of turning public spaces from a boring reflection of a consumer culture into sites of stimuli and play.[3] Following the footsteps of these predecessors, our wish here is to highlight the biases inherent to the spatial perception of the OPT, with the hope that their partial understanding can be remedied.

Many of the studies that followed Lynch and Debord demonstrated how social groups differ in the ways they discern common spaces. A classic among them is the work of Gould and White (1986, originally published in 1974). This "multiple rendering of space" was also considered in the context of the Palestinian–Israeli conflict, when Waterman (1980) compared the paths which Arabs and Jews traverse while walking through Acre, a "mixed" city. A contemporary example comes from Belgium, where political scientists wished to measure territorial conflict by comparing mental maps of French-speaking and Dutch-speaking citizens (Reuchamps *et al.* 2014). Mental map studies also shed light on the preservation of spatial segregation, originally set during the colonial era and persisting in the form of socio-economic divisions after the demise of the empire, as Smiley (2013) demonstrated in present-day Dar es Salaam. Such studies go beyond outlining social tensions, unequal power relations, and a diverging use of space. They reveal implicit spatial sensitivities that guide people, yet of which they are often unaware. Our work follows this path, with an attempt to explore spatial perceptions of the Palestinian–Israeli Conflict fixed in people's minds yet often undiscussed.

Central to our paper is the concept of deterritorialization, reflecting the fact that people find it difficult to grasp the spatial contours of the conflict's territory. While Deleuze and Guattari (1987, pp. 431–437) introduced issues of deterritorialization back in the 1970s, alluding to a new mediated relationship between people/labor and land, the term was later invoked to describe general trends in the deteriorating ties between cultural forms and specific territories. Deterritorialization was used to discuss relationships between nation-states and their diasporas (Basch *et al.* 1994) and the emergence of virtual territories such as the internet, that have the power to create deterritorialized protest (Capling and Nossal 2001). Deterritorialization, argued Appadurai (1990), is a central force in our globalized world. One of its traits is that state and nation "are at each's throat" and "there is a battle of the imagination, with state and nation seeking to cannibalize one another" (1990, p. 304).

The mental maps of Palestine–Israel produced by our subjects, and the interviews that accompany them, illustrate this rupture between state, nation, and land. This rift may be bound to global trends yet takes on unique expressions in different contexts. Deterritorialization in the OPT is a product of multiple spatial arrangements, especially those developed during the five-decade Israeli policy of non-annexation, Jewish settlement, and Palestinian dispossession.

A concise recent history

The Israeli–Palestinian Conflict evolved during the late Ottoman Rule, when European Jews arrived under the banner of Zionism in the Holy Land with the aim of establishing a state for the Jewish people (Ram 1999; Shafir 1999; Smith 2010). This project soon led to the dispossession of local Arab populations and conflicts between the groups erupted early on, gradually intensifying (Khalidi 1997; Farsoun and Aruri 2006).

By the end of World War I, the United Kingdom captured the area between the Jordan River and the Mediterranean, and shortly thereafter, the League of Nations granted it a mandate over Palestine. The country was given an official Arabic name, *Filastin*, and a Hebrew name, *Eretz Yisrael*, both names resonating a history that goes back thousands of years. When three decades later Britain requested to end its mandate over Palestine, the United Nations prepared a partition plan, envisioning two future states – an Arab and a Jewish one.[4] However, a year-long war broke out in late 1947, and the Zionist forces prevailed, establishing a Jewish state at the expense of the Arab population of Palestine (Morris 1987, Smith 2010).

The new state of Israel did not extend over all of British Palestine. A part of Eastern Palestine came under Jordanian rule and was named the West Bank.[5] In addition, Egypt won control over a part of the southwestern edge of the country, that came to be known as the Gaza Strip.[6] Both these areas together comprise 22% of mandate Palestine's territory (Figure 1).

During the 1947–1948 War, the majority of Arabs living in what would become Israel were either forced to flee or expelled outright and were not granted permission to return (Morris 1987). While mandate Palestine used to be home to roughly 1.3 million Arabs, more than 700,000 became refugees (UNRWA 2007), first being dispersed to neighboring Arab countries and then settled in camps built by a special United Nations division established for the Palestinians – The United Nations Relief and Work Agency (UNRWA).

During the June 1967 Six-Day War, Israel occupied the remainder of mandate Palestine – namely the West Bank and the Gaza Strip, as well as the Sinai Peninsula, formerly under Egyptian rule, and the Golan Heights, formerly part of Syria (Smith 2010). The Palestinians who resided in Jordan's West Bank and in Egypt's Gaza Strip until 1967 now came under Israeli

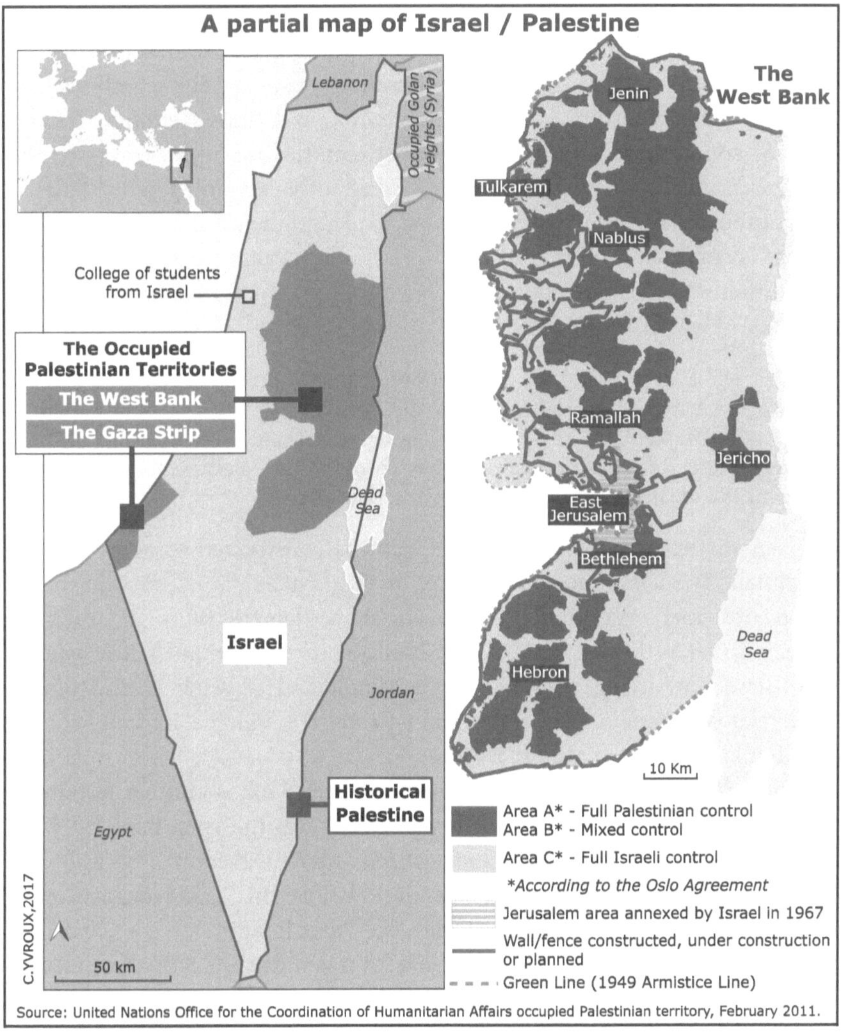

Figure 1. A partial map of Israel/Palestine.

domination, or escaped and became refugees for the second time. Areas adjacent to Jerusalem were unilaterally annexed by Israel shortly after this war (Lustick 1997; Rempel 1997). According to Israel's judicial system, those Palestinians residing in the annexed areas hold the status of "inhabitants", which grants them limited political and social privileges (Kretzmer 1984).

In the 1990s, following peace initiatives and the signing of the Oslo Accords, the OPT were subdivided into three areas: Zone A, under the control of the Palestinians; Zone B, militarily Israeli but under Palestinian civil control; and Zone C, comprising the vast area of the Occupied Territory, under full Israeli control. This division persists to this day, although it was supposed to have been a temporary one leading to a Palestinian State.

Beyond the fragmented character of this territory, there are two main national populations who reside inside it, holding very different statuses: the Palestinians, who are subordinate to military rule, and the Jewish settlers, who hold full citizenship. Although these two populations seemingly inhabit the same space, there is little similarity between the areas they traverse. The Jews live in state-sponsored settlements, constantly expanding, while Palestinians cannot move freely within the West Bank and their infrastructure development is curtailed (Ophir *et al.* 2009). Spatially speaking, according to Levy (2008), there is a "superimposition of two topographies, including two road networks". Levy further notes:

> The Israeli layer is clearly the dominant one: it appropriates the existing road network whenever necessary and has the best accessibility. The Palestinian layer is incomplete, full of stops and detours, but locally enjoys connectivity through underground passageways that connect the Palestinian areas geographically separated by Israeli de-construction. (Levy 2008)

The "Jewish topography" of the West Bank is constructed so as to prevent Palestinians from free movement while giving Israelis the impression that it is part of Israel (Weizman 2012).[7] Jewish settlement blurs the division between Israel proper and the West Bank, with more than half a million Israeli-Jews now living within the West Bank and crossing in and out of Israel relatively freely.[8] Yet, Palestinian-Arabs are still the majority in the West Bank, numbering almost 2.7 million.[9]

The mere disagreement over naming the West Bank is another testament to its multiple identities. Under Jordanian rule, it was the West Bank (*al-Diffa al-Gharbiyya* in Arabic and *HaGada HaMa'aravit* in Hebrew), occasionally shortened to the Bank (*al-Diffa/HaGada*). While this name persists, it is also known as the Occupied Territories, the Palestinian Territories, the Territories, or Palestine. Israel's official name for it is Judea and Samaria, creating the impression of a direct link from the Biblical Era to this day. The Israeli government decided back in 1967 to omit from official maps the "Green Line", namely the de facto border of Israel until 1967, based on the 1949 armistice agreement line, thus further blurring the divide between Israel proper and its military-ruled areas (Leuenberger and Schnell 2010). In 2002, the Israeli government began building a separation wall/fence between Israel and the West Bank, yet much of it does not run along the Green Line (Figure 1), and portions of the West Bank are not separated from Israel proper by a wall or by a fence.[10] The population of the State of Israel proper, meanwhile, consists of 20% Arab-Palestinian citizens, some of them maintaining social, cultural, and economic bonds to the OPT (Forte 2001).

The situation between Israel and the Gaza Strip is different. In 2005, Israel withdrew unilaterally from the Strip and removed its settlements there. The

Strip was surrounded by a "smart" fence and since Hamas came to power in 2007, it has been under siege, with Israel monitoring all comings and goings, and much of the movement in and out done clandestinely, through underground tunnels to Egypt.[11] It is significant to note that the Gaza Strip is almost entirely closed off from the outside world. In our study, we ask how these two places – the West Bank and the Gaza Strip, territories ruled by a set of harsh and uncommon arrangements – are imagined and understood by our participants.

Methodology

We approach this study first from the distant outsiders' point of view, from the perspective of those who do not experience the conflict first hand. We study the French because while theirs is the perspective of "outsiders", they are also known to be passionate about the Palestinian–Israeli conflict. Several explanations have been suggested for this, be they based on political affiliation, identity issues, or, as some say, "community disposition" (Sieffert 2004, Gresh 2010). A possible reason is the importance of the Jewish population in France as well as that of immigrants from the Maghreb, both considered to have a rather sympathetic position in relation to this conflict, the former to Israel and the latter to the Palestinians. A political–historical perspective would add that questions of colonialism, imperialism, and international law are cardinal to the French tradition and ethos. Thus, it sometimes seems like everyone in France has an opinion regarding the conflict.

This was a promising starting point for Yvroux to choose a French population for this study, conducted among undergraduate students of geography in May 2016. It came as a follow-up to a much larger survey on the topic conducted 7 years earlier (Yvroux 2012). While the 2016 survey placed an emphasis on qualitative analysis, the previous one was more quantitative. When presenting our results, we will rely primarily on the 2016 study yet will refer to some of the findings obtained from the 2009 study, among 221 students.

Geography students were chosen both because they are assumed to be interested in geopolitical issues and because they have had experience with mapping. In that sense, they are probably better equipped to carry out this exercise compared to other French populations. Yvroux collected sketch maps and interviewed 15 students, 9 men and 6 women, within the age range of 19–28 years. At the end of a course, respondents were asked to volunteer as participants. Two sketch mapping exercises were requested from students who have learned about mental maps; they knew that they should register anything that comes to mind regarding the territory. The first exercise was to fill in a fixed frame which contained the borders of Mandate Palestine and the second was to complete a regional map, which had the borders of the different political territories – Gaza, the West Bank, and neighboring countries (sketch map number 2 on Figure 2). For each of these exercises, the same instructions were given: "Here is the conflict's territory. Fill in this map with all the elements you know". Presenting the students with a fixed frame, rather than no frame at all, was essential to reach workable results. Still, the task posed a challenge for the students, because of the inherent difficulty of drawing a map and also because the students felt that they were being tested, which added to their discomfort. Other studies of mental maps pointed towards such reactions (Smiley 2013, p. 226). To enhance our understanding of the mental map sketches, we included an interview with open-ended questions, discussing the conflict's location, the claims of each adversary, and the understanding of the following terms: The Gaza Strip, settlements, wall, Jerusalem, refugees, and the West Bank.

The findings from Israel are also based on a study carried out in 2016 and draw on a larger study conducted between 2009 and 2011 (Ben-Ze'ev 2012 and 2015). The 15 students (4 men and 11 women) who participated in the 2016 study were all undergraduate students in Behavioral Sciences at a college in the center of Israel. All were in their early to mid-twenties, apart from two men who were in their thirties.[12] Four of the 15 were Arab-Palestinian while the others were Jewish, all citizens of Israel.[13] Unlike the French students, in the Israeli case study, we simply opted for the "average student", for those who experience the territory on a daily basis. We did

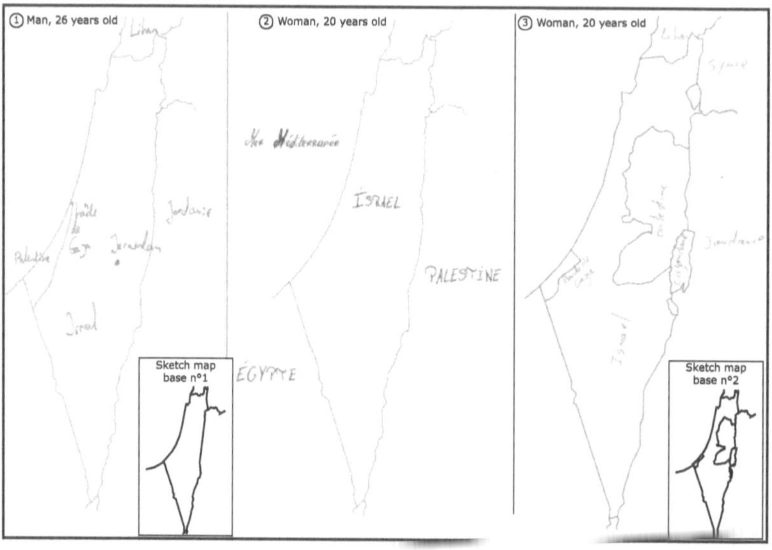

Figure 2. Examples of students' mental maps, France.

not wish to have experts here but rather laypeople. These first-year students were obliged to participate in 10 hours of research and were free to choose which study; this project was one of their first choices. There were more volunteers than we needed, so the first to apply were the ones we selected. Yet because we were exceptionally curious about the Palestinian-Arab perspective, we included all four Palestinian-Arab students who volunteered. Our study could have gained further insight had we included Palestinians living inside the Occupied Territories. However, in this case, we decided to focus our efforts on those who live outside the territory, both near and far.

Our study in Israel combined two main methods: one was the request to sketch a map of their "country", and indicate on it borders, cities and towns, regions, nearby states, and their capital cities, as well as the Green Line. This was followed by interviews, through which other topics were explored, such as why certain information was on the map while other was missing and where and how the participants gathered their information. We discussed topics such as family trips, school excursions, and memorable encounters with the borders. These interviews revealed parts of the process through which they acquired the spatial knowledge as depicted in their mental maps. Our intention was to determine the correspondents' subjective reality and to get a better understanding of their perception of the country's territory, while treating their representations neither as true nor false (Bailly 1992).

Mediated spaces: the mental maps of students in France

The cognitive psychologist Barbara Tversky argued that mental representations of spatial knowledge are necessarily distorted, fragmented, and incomplete (1993, pp. 14–24). These distortions are of great interest to us, as they tell us something about the unique perceptions of the conflict among different social groups. The analysis of the French students' representations revealed fragmentation and lacunae. The students have bits and pieces of knowledge, while on some issues they confess to knowing nothing, leaving empty regions on the map. At the same time, because they do try to create a spatial image of the conflict, they labeled place names almost randomly on the map. Clearly, these students have difficulties conceptualizing the territorial dimensions of the conflict. In fact, it seems deterritorialized, because it is no longer clear where it takes place.

The main result of our inquiry was an insight into the prevailing confusion over the nature of "the Palestinian Territories", "Palestine", "the West Bank", and "the Gaza Strip". The students did not use the term "Occupied Territories" for a separate entity, either on the maps or in the interviews. They seemed to imagine that there is just one territory for Palestinians and it usually is called Palestine. This may be the outcome of the expression "the

Israeli–Palestinian Conflict", which they assumed to be between two territorial entities: Israel and Palestine. When the students were asked about the location of this conflict, they answered "Israel and Palestine", sometimes omitting Palestine. In their understanding, the confrontation takes place on a border area, but few of them tried to map this vision (Figure 2).

For some students, the Gaza Strip is where the border is, described by one participant as "a neighboring region between Israel and Palestine, a serious conflict area" (woman, age 20). For another student, the confrontation takes place on the "Israeli–Palestinian border" and the Gaza Strip is an additional matter, because it is "claimed by Palestinians, appropriated by Israelis, [and] also initiating the conflict" (man, age 20). The second map to fill in (Figure 2, sketch map base #2) confirms the dissociation between the Gaza Strip, the West Bank, and Palestine. Thus, some students marked these three names on the map, with Palestine in place of the West Bank and the West Bank instead of Jordan or the Dead Sea (Figure 2, sketch map #3).

If using the term "Palestine" seemed spontaneous for the majority of the students in the open-ended questions, the idea of a Palestinian territory still remained vague. Nevertheless, there are two apparent patterns. The first is that the Gaza Strip is seen as connected to the conflict and its territorial issues but not necessarily to the Palestinians. The second is that the West Bank territory remains an entirely undefined concept.

The Gaza Strip was the most common element in the map sketches. Most of the time, it appeared as a strip in the southwestern area of the map. Most of the students connected the Gaza Strip with violent connotations: A "place where the conflict takes place" (woman, age 19), "attack, war, rocket launcher" (man, age 28), "war, conflict territory" (man, age 20). These were often the only features that they mentioned, without offering information either on the inhabitants' identity or on the concrete issues that are at stake. Some students, nevertheless, did have more specific insights: "Palestinians are living there and part of them deeply claim the territory as theirs and perpetrate terrorist acts. In return, Israel bombs them, often leading to civilian casualties" (man, age 26). At the same time, others reflected on their lack of knowledge: "I don't know much about the Gaza Strip. The media speaks about it a lot but they don't explain what it is" (man, age 26).

The incomprehension of the West Bank Territory remained a dominant feature of the students' representations. In French, the "West Bank" is called "Cisjordanie," literally translated as "this side of Jordan." Only once did the term pop up spontaneously. When the students were asked about this territory in particular, they explained that it is "a country in the conflict area" (woman, age 20); "a country involved, a neighbor" (woman, age 20); "it seems that it's an old country, at least a region" (woman, age 20).

If the students linked the West Bank with the Palestinians, it was often in relation to the refugee issue: "A reception area, given by the king, to welcome the Palestinian migrants driven out by Israel" (man, age 24) or "a place where the refugees go" (woman, age 19), or sometimes: "helps Palestine non-officially, openly anti-Israeli" (man, age 23); "an ally of Hamas" (man, age 23). The formulation which was given by one of the interviewees summarizes well the general feeling: "A bordering country hosting refugees, which places itself in conflict with the state of Israel. Even if I don't know exactly its role – this country often appears in the Israeli–Palestinian conflict" (woman, age 20).

Students may have strong and precise ideas (occasionally wrong) of the issues pertaining to the conflict, but insufficient spatial and cartographic knowledge; their sketched maps are scant. However, the students were well aware of the fact that the conflict is about land issues and when asked about the Palestinians' claim, they answered that it concerned "a part of the territory which belongs to Israel" (woman, age 28); "some land" (man, age 20); or "they want to get back a territory which formerly belonged to them – the whole Israel or Gaza Strip, I don't remember – but it was assigned to Jews after the Second World War" (woman, age 20).

These findings may be understood within the territorial schemata available to French students. They live in a state where national and local boundaries are set and recognized, and in a relatively peaceful environment. Within this geographical and geopolitical culture, it is difficult for them to imagine other forms of sovereignty – split sovereignty, lack of sovereignty, multiple sovereignty – and thus to comprehend a particularly complicated territoriality such as that of the West Bank. It does not fit their common knowledge. Moreover, all these students were born while the peace process supposedly had begun and lack a longer historical perspective. In France, the conflict in the Middle East receives high media coverage, but it does not allow for a full picture. French journalists use different words and phrases such as Gaza, West Bank, Palestine, Palestinian Territories, Occupied Territories, Occupied Arab Territories, the Territories, Palestinian entity, and Territory of Judea-Samaria (Yvroux 2012). With so many names, it is not surprising that the French students do not have a clear vision of the territory.

Veiled territory: the mental maps of students in Israel

While the French students gathered their impressions of the West Bank and Gaza mainly through the media and social media, education, and books, students in Israel live nearby, and some think of these territories as part of their country. In the exercise, they were asked to draw their country – *ha'aretz* in Hebrew and *al-bilad* in Arabic – on a blank page. Geographical lacunae and mistakes were naturally part of their maps, allowing us a glimpse into what *their* country consists of.

There was a similarity between the French and the Israeli maps. In both cases, the Gaza Strip was a clearer territorial entity than the West Bank. The Gaza strip was more often indicated on the Israeli students' maps when compared to the appearance of the West Bank and often, the former was sketched as larger in proportions than it really is. Yet while the Gaza Strip appeared in many of the maps, its location sometimes "drifted" (Figure 3(1)).

In contrast to the prominence of the Gaza Strip, only half of the students were able to draw the perimeter of the West Bank. In other words, on half of the maps, the West Bank area was represented as integral to "the country". Even when it was outlined, it seemed that the area posed a terminological impasse; it was simply not named. This is perhaps not surprising, as there are many forms of the country to which students are exposed in school (Ben-Ze'ev 2015). Of the 15 students interviewed in 2016, one named it "the territories", one named it "Judea and Samaria", and one, a Palestinian from Haifa, wrote "the West Bank" in English. On the other maps where it appeared, it was nameless, though other regions such as the Negev or the Galilee were named.

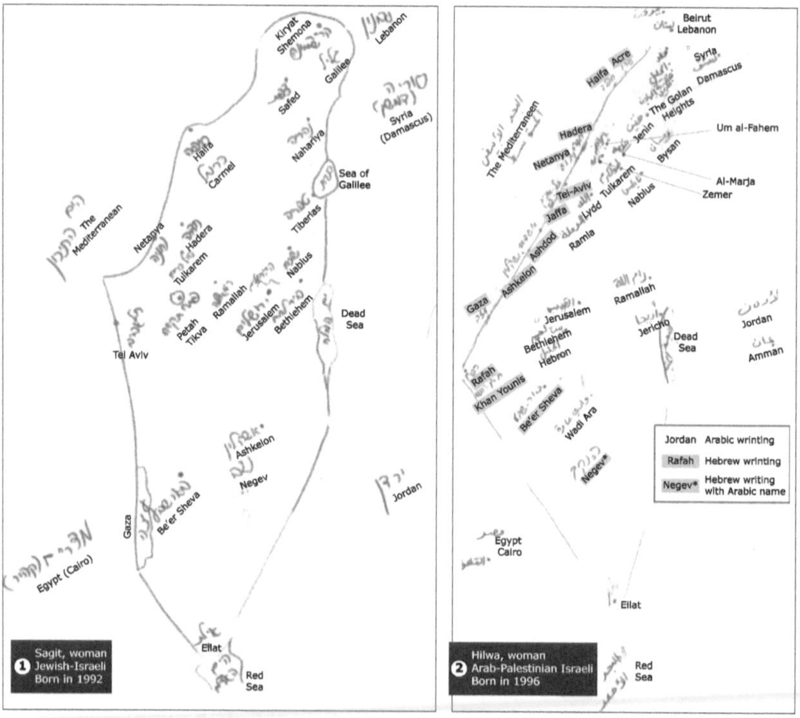

Figure 3. Examples of students' mental maps, Israel.

Another feature, which demonstrated the misunderstanding of the West Bank environment, became apparent when most of the students could not locate its main towns. When being asked to do so, they reacted by either scattering the names randomly or abstaining from indicating them. In one example, Jenin and Hebron were missing, Ramallah was set west of Jerusalem, rather than north, and Bethlehem was placed east of Jerusalem, rather than south (see Figure 3(1)).

If the West Bank area was often incorporated into the boundaries of "the country" and lacked a perimeter, the country as a whole was usually sketched with clear outer lines, like a disconnected island (Figure 3(1)). What lay beyond this island was unknown: Almost no one was able to draw the borders of neighboring states and the seam-line between these neighboring states and the Mediterranean. The neighboring Arab states were terra incognita, including Jordan and Egypt, with whom Israel has open land crossings.

The students re-created their country as an enclave, a floating one with a clear perimeter, surrounded by unknown neighboring states and the Mediterranean Sea. This fortified enclave contained "internal enclaves", or unknown "white patches", the main one being the West Bank. In a classic piece titled *Seeing is Believing*, Dundes (1972) argued that sight is perceived as the prime faculty in western cultures. Yet, seeing is not always believing. In what follows, we explore how eyesight is applied in a selective manner, as becomes evident through the students' descriptions of their encounters inside the West Bank.

Jewish-Israeli students' reflections on visiting the West Bank

Many of the Jewish-Israeli interviewees declared that they avoid the West Bank. Even though it is almost impossible to live in Israel without entering it, they were simply unaware of having done that. We chose those who were aware of these visits in order to consider their narratives of the encounter. One interviewee was Gaya, born in 1994, who grew up in a middle-class town in the center of the country, in a Zionist religious Ashkenazi family associated with the settlers' movement. Gaya noted that she has friends in settlements and that her brother studied inside the West Bank. Therefore, unlike secular students, who often claimed to abstain from going into the West Bank, Gaya occasionally did. Her descriptions of these visits demonstrate that they were not experienced as part of the daily routine.[14] Gaya:

> Once we were taken to Rachel's tomb. It's in Bethlehem, right? They took us there once and it was like a secret operation. "Let's do it quickly and get out of here because it's a little dangerous". We were brought there and told that our time is limited: "Go in, pray, and then we should leave". We also went to

Hebron and they kept reminding us that it's dangerous, that we should stick to the group and stay close to each other. During national service we were also taken to Hebron because it's close to Jerusalem.

What do you remember?

Hebron looked deserted; no one was on the street. We were there at this place, what's it called, The Cave of the Patriarchs. I was there twice in my life. Once it was the eve of a holiday and it was crowded with [Jewish] people and therefore wasn't disturbing; and then again it was a trip during the daytime, and it looked empty and less of a touristic place, unlike what I had remembered from my previous trip. In Hebron I remember half demolished houses, a deserted street, and soldiers here and there. One hurries to leave. You finish what you do and wait at the bus stop and there were a couple of Arabs and there is real anxiety. We didn't have a guard. We waited for the bus and the bus ride itself is frightening. You could say that you put your head down and try to disappear, avoiding eye contact with your surroundings. I'm sure it's even more frightening these days.

Gaya's description tells us something about the passivity she experienced when going into the West Bank ("they took us there"); perhaps, it was not something she would initiate herself. Hebron, for Gaya, was transformed at nighttime, during a holiday; there was a shifting sense of belonging. When Jews surrounded her, she seemed unaware of the Arab setting. But her daytime visit was daunting, and as she rode the bus back, she chose to "disappear", avoiding her surrounding and hoping not to be seen. Two points are evident: One is that the very same place inside the West Bank can transform from "feeling safe" to "feeling threatening"; second is that invisibility is described as an almost voluntary act and it works in two directions – wishing neither to see nor to be seen.

Our second example adds another dimension to the Jewish-Israeli puzzlement experienced inside the West Bank. Nadav, born in 1977, grew up in Ashkelon, a Mediterranean town bordering with the Gaza Strip. At the time of the interview, he was a youth counselor at a boarding school, telling us how he and his students were taken (again, passivity) on an educational excursion named "Israeli Journey".[15] This is how he described part of this tour:

I was with my students in Gush Etzion [a settlement bloc South of Jerusalem] and it surprised me. As part of this one-week journey, we were taken to a big tent in Gush Etzion. It is called Kefar Ayalim. And it bothered me. We crossed a checkpoint and reached a high point at Gush Etzion. "How did we get here?" [he says, laughing]. There were guards. It felt OK. The students were not aware of it. You don't choose to share such information [with them]; you don't want them to be anxious (*beharadot*).

But a week or two later there were troubles at that junction – a stabbing or a car-ramming attack or maybe shooting. It sounds like another country. It doesn't

> sound like a place that you visit on a daily basis. You didn't really mean to get there.

Nadav was uneasy with the trivialization of the visit ("how did we get here?"). At the same time, he was the single student who registered it on the map-sketch as "Judea and Samaria", thus adopting the official Israeli name. Yet he wished to express dissatisfaction with the concealment the danger looming in that place ("a week or two later there were troubles at that junction – a stabbing or a car-ramming attack"). He felt that he was in "another country", with rules of its own. Yet at the same time, he participated in the process that obscured the ambiguous nature of the place, both in accepting the official Israeli name and in abstaining from clarifying to his students where they were.

Palestinian-Arab–Israeli reflections on visiting the West Bank

The two separate West Bank topographies described earlier by Levy – a Palestinian and a Jewish one – transpire through the experiences offered by Gaya and Nadav, both Jewish-Israeli. However, Palestinian-Arab students, citizens of Israel, described a different kind of experience. They ventured into the West Bank for a variety of reasons, among them, to meet their friends and family – Palestinians who are under military rule and cannot enter Israel. While within Israel, Palestinian-Arabs are treated as second rate citizens (Yiftachel 1999), the West Bank has much to offer them – Arabic is the lingua franca, they have professional ties and relatives, and services and shopping are cheaper. Yet despite the advantages, going to the OPT is also accompanied by discomfort, often stemming from the experience of border crossing, both social and spatial.

Hilwa is a Palestinian-Israeli woman born in 1996 in a village near the "green line" (see Figure 3(2) for her map). When asked about this border, she said: "It is right near me. I can go to Jabel al-Jaber and I'll be right on the green border". This curious slip of the tongue indicates that for Hilwa the green line is a recognized border. She added: "I can see the army cars passing. The army would pass with its armed vehicles [*dababeh*]."[16]

Hilwa's family regularly goes into the West Bank for shopping and medical care.[17] Her sisters, she noted, like going there because "it's cheaper, and they like dresses". In the interview she described her last visit to Tulkarem, inside the West Bank:

> I go there often … It takes us a maximum of half an hour. There's a checkpoint. It is a little disturbing. You feel a little foreign. You feel as if suddenly you are in a different place but it's not as if you have to be there.

> There are always many children sitting around there. But [she hesitates], in there [abstaining from giving the place a name] you don't feel it so much [meaning, it's

> a little different]. They [OPT Palestinians] can tell if you are from here or there because of our clothes. They see. There's a difference. We don't dress as they do. Even the head-cover is not the same as theirs. And the makeup is different.

Hilwa's narrative tries to bypass a set of sensitive issues: The checkpoint crossing, manned by Israeli-Jewish soldiers, who can decide whether the family will enter the OPT; feeling that she crosses a threshold into a new environment which is both inviting and triggering discomfort; meeting the children in Tulkarem who are probably begging (and exposing the class differences); being aware of the different dress code, indicating that she is foreign. Yet, despite all those, she and her family are attracted to Tulkarem.

At the same time, she noted, they abstain from visiting other places inside the West Bank. Elsewhere in the interview, she clarified that while her family travels to Tulkarem regularly, her father does not permit them to go to East Jerusalem. There, he argues, it is far more dangerous as they can be mistakenly taken for Palestinians who do not hold Israeli citizenship and hence become exposed to the violence of the military forces. As we see, the West Bank in Hilwa's narrative is not a singular place; rather, it divided into "safe" and "non-safe" places.

These subtleties of the hospitable and the non-hospitable, those places to be entered and to be avoided, guide the movements in a conflicted Palestine/Israel, and are especially evident in the West Bank. Gaya described a confusing trip whereby she entered a Jewish-Israeli "enclave" – the Cave of the Patriarchs in Hebron – which is encircled by what she experienced as an Arab-Palestinian "enclave" – the town of Hebron. Hebron itself is part of the larger militarily occupied West Bank, ruled by Israel. These enclaves within enclaves are like concentric circles, each experienced as different in its national identity. Moreover, the interviewees described their intricate knowledge of different codes that operate within the OPT – Tulkarem and East Jerusalem are not the same from Hilwa's family point of view, and that affects their decisions (on whether they will go there or not). Another point, the very same place can change its "nature", such as Hebron for Gaya between her first and second visit, or the junction described by Nadav, where a stabbing or a car-ramming incident was carried out shortly after his visit. Fourth, Nadav adopts the Hebrew name of Judea and Samaria, supposedly assuming the Jewish nature of this place, yet is anxious when he discovers that he is there with his students. The nature of this territory shifts; there is no consensus over its character, its continuity, or its mere existence as a single entity.

Deterritorialization and reterritorialization

Processes of deterritorialization manifest themselves in the Palestine–Israel Conflict and specifically in the West Bank. The sketchy maps of the French students indicate that they can barely relate to the territorial dimensions of the conflict. This is a lesson concerning the diffused image of this high-profile conflict. Even in the minds of an interested audience, it does not have a graspable spatiality. Moreover, the mental maps that these students draw are perhaps so lacking because they are based on official maps which themselves offer a "truncated" vision of reality (Pinheiro 1998; Battersby and Montello 2009).

Yet the West Bank is also a riddle for those who know it, if only partially. Students from Israel sketch it as part of their country yet its content remains unknown. A valuable future study would be to collect mental maps from Palestinian students inside the OPT and interview them. Deterritorialization will possibly remain an issue in their case too, as impediments to free movement and the presence of "no-enter" settlements are integral to their lives and may also fragment their mental map of their territory.

Our focus here was on the psychological dimension of distance. As we have seen, "over there" can be next door. There are at least two "layers" of the West Bank territory, experienced as either totally apart or enmeshed in one another. Hence, for the students from Israel, imagining their country, including its militarily occupied parts, depends not only on their physical contact but also on what they are capable of witnessing and acknowledging.

The territoriality of the West Bank and the Gaza Strip does not fit into the familiar classic idea of a nation-state. Gaza is closed off, an unknown place – a white patch, a black hole. In contrast, the West Bank is a puzzle – it often has no evident boundaries, the unknown green line, and one can enter it without noticing. At the same time, one encounters in its midst roadblocks, high fences, and walls. It is a place of multiple topographies which can easily create a baffling experience.

The fact that Palestine/Israel, and especially the West Bank, appear to be deterritorialized, is also the result of a process that serves Israel's policy of deepening the occupation. If the political situation is presented as disconnected from its territorial dimension, or appears to be extremely complicated, it is only natural to assume that there is no solution; it is part of the management of perception. Much of the political and media discourse, both inside Palestine–Israel and outside of it, highlights the peace process and a two-state solution. However, the perceptions we explored suggest that there are many and different political entities within Palestine/Israel, particularly within the OPT of the West Bank. If a two-state solution is still an option, Israel's policies of fragmentation and concealment should be taken into account.

Since we tried to highlight the filters affecting the ways our students imagine the conflict, we should do the same for ourselves, the authors of this paper. First, we analyzed the students' representations as their reality, as a world created by multiple and contradictory discourses. Second, as scholars, we tended to compare their mental maps to academic, official, and scientific knowledge, and we were inclined to be judgmental about "inaccuracies", deeming certain details as right or wrong. Third, we wished to bring forth this conflict's spatiality and the gaps in its understanding. Trying to describe the situation in the fairest manner is not an easy task. When we write of the West Bank, should we treat it as a territory, an entity, a multi-layered place? Since it has become so fragmented, perhaps it no longer makes sense to describe it as a single entity. How should it be represented on the map? When we chose to draw a map of the West Bank with an outer line and a different color, did we not invent a spatiality that is far from being so in the minds of many people? Scientific maps should be considered as representations of one type of discourse, while our students' maps communicate a message that does not come across in standard cartographic representations, yet may be as important. So many meanings, layers, and names have piled up on this disputed territory, and especially on the West Bank, that it appears to be buried underneath them. Can a real territory emerge from this?

Notes

1. The earliest experiments of mental maps, given the name cognitive maps, were carried out on rats by psychologist Edward Tolman (1948).
2. Lynch (1960) was seeking to decipher which spatial elements are legible and what he named "imageable," namely, bearing a familiar image.
3. Guy Debord, a leading thinker of the Situationist International, was influential in the emergence of the 1960s students' revolution in France. One of Debord's suggestions for improvement was to build public playgrounds for adults. See the film *The Situationist International 1957-1972*, https://www.youtube.com/watch?v=2SvdWk8zRrI [Accessed 14 August 2017].
4. This partition plan led to the UN resolution 181. See http://www.un.org/Depts/dpi/palestine/ch2.pdf.
5. The dimensions of the West Bank are roughly 120 km long and 35 km wide, covering 6000 square km. The name evolved in contrast to Jordan's larger stretch east of the Jordan River.
6. The Gaza Strip is roughly 40 km long and between 6 and 12 km wide, covering 360 square km.
7. The free flow between the Jewish settlements of the West Bank and Israel proper is carried out in a variety of ways: by blurring the 1949 armistice agreement boundary known as the Green Line, including the establishment of Jewish settlements in it; by moving checkpoints away from the Green Line; and placing signs that warn Israeli Jews only when they leave area C and enter Palestinian villages and towns, and not when entering Israeli-controlled West Bank territory.
8. In French it is Cisjordanie, literally meaning "the same side of Jordan", which probably evolved in contrast to Trans-Jordan – beyond the Jordan River.
9. The number given for the West Bank Palestinian population by the U.S. Central Intelligence Agency is 2,697,687 (July 2016 est.). See https://www.cia.gov/library/publications/the-world-factbook/geos/we.html [Accessed 28 January 2017].

10. This wall is planned to be 712 km long, twice the length of the Green Line (OCHA Opt, 2013).
11. See, for example, http://content.time.com/time/photogallery/0,29307,1931308_1969745,00.html [Accessed 14 August 2017].
12. The sample reflects the higher percentage of women in this department.
13. Palestinian-Arabs are roughly 20% of Israel's population, yet their ratio in this college was lower.
14. Religious women in Israel can substitute their mandatory army service with "national service" (*sherut leumi*), or volunteering in civilian institutions. Gaya said that her travels into the OPT were part of her national service.
15. The Israeli Journey is a semi-independent educational program linked to the Ministry of Education. Among other activities it offers a 6-day trip for 11th grade students through different places in Israel and the OPT. See their site: http://masaisraeli.co.il/languages/english/about-us/ [Accessed 8 August 2017].
16. The interview was conducted in Hebrew but was interspersed with Arabic words such as *dababeh*, Arabic for an armed vehicle.
17. Hilwa told a story about a doctor from the OPT who solved an allergy problem she suffered from, which doctors in Israel failed to diagnose.

Acknowledgements

The authors would like to thank Janne Holmén, Norbert Götz, Steven Schnell, and the anonymous reviewers for their useful comments.

Disclosure statement

No potential conflict of interest was reported by the authors.

Funding

This study was supported by the Ruppin Academic Center, the Harry S. Truman Research Institute for the Advancement of Peace, Hebrew University, and the GRED laboratory of Paul Valéry University.

References

Appadurai, A., 1990. Disjuncture and difference in the global cultural economy. *Theory, Culture & Society*, 7, 295–310.

Bailly, A., 1992. Les représentations en géographie. *In*: A. Bailly *et al.*, eds. *Encyclopédie de la géographie*. Paris: Economica, 372–383.

Basch, L., Glick-Schiller, N., and Szanton Blanc, C., 1994. *Nations unbound: transnational projects, postcolonial predicaments and de-territorialized nation-states*. London: Taylor and Francis.

Battersby, S.B. and Montello, D.R., 2009. Area estimation of world regions and the projection of the Global-Scale Cognitive Map. *Annals of the Association of American Geographers*, 99 (2), 273–291.

Ben-Ze'ev, E., 2012. Mental maps and spatial perceptions: the fragmentation of Israel-Palestine. *In*: L. Roberts, ed., *Mapping cultures: place, practice, performance*. London: Palgrave Macmillan, 237–259.

Ben-Ze'ev, E., 2015. Blurring the geobody: mental maps of Israel/Palestine. *The Middle East Journal*, 69 (2), 237–254.

Capling, A. and Nossal, K.R., 2001. Death of distance or tyranny of distance? The internet, deterritorialization, and the anti-globalization movement in Australia. *The Pacific Review*, 14 (3), 443–465.

Deleuze, G. and Guattari, F. 1987 [originally published 1980]. *A thousand plateaus: capitalism and chizophrenia*. Minneapolis, MN: University of Minnesota Press.

Dundes, A., 1972. Seeing is believing. *Natural History*, 81 (5), 8, 10–12, 86–87.

Farsoun, S.K. and Aruri, N.H., 2006. *Palestine and the Palestinians: a social and political history*. 2nd ed. Boulder, CO: Westview Press.

Forte, E., 2001. Shopping in Jenin: women, homes and political persons in the Galilee. *City & Society*, 13 (2), 211–243.

Gresh, A., 2010. *De quoi la Palestine est-elle le nom?* Paris: Les liens qui libèrent.

Gould, P. and White, R., 1986 [originally published 1974]. *Mental maps*. 2nd ed. Boston: Pelican Books.

Knabb, K., ed. and trans., 1981. *Situationist international anthology*. Berkeley, CA: The Bureau of Public Secrets.

Khalidi, R., 1997. *Palestinian identity: the construction of modern national consciousness*. New York, NY: Columbia University Press.

Kitchin, R. and Blades, M., 2002. *The cognition of geographic space*. London: Tauris.

Kretzmer, D., 1984. *Israel and the West Bank: legal issues*. Jerusalem: West Bank Data Base Project.

Leuenberger, C. and Schnell, I., 2010. The politics of maps: constructing national territories in Israel. *Social Studies of Science*, 40 (6), 803–842.

Levy, J., 2008. Topologie furtive [online]. *EspacesTemps.net*. Available from: http://www.espacestemps.net/articles/topologie-furtive/ [Accessed 17 August 2017].

Lynch, K., 1960. *The image of the city*. Cambridge, MA: MIT Press.

Lustick, I., 1997. Has Israel annexed East Jerusalem? *Middle East Policy*, 5 (1), 34–45.

Morris, B., 1987. *The birth of the Palestinian refugee problem 1947-1949*. Cambridge: Cambridge University Press.

OCHA, 2013. *The humanitarian impact of the barrier* [online]. OCHA Opt. Fact sheet. Available from: https://www.ochaopt.org/documents/ocha_opt_barrier_factsheet_july_2013_english.pdf [Accessed 16 February 2017].

Ophir, A., Givoni, M., and Hanafi, S., 2009. *The power of inclusive exclusion: anatomy of Israeli rule in the Occupied Palestinian Territories*. New York, NY: Zone Books.

Pinheiro, J.Q., 1998. Determinants of cognitive maps of the world as expressed in sketch maps. *Journal of Environmental Psychology*, 18 (3), 321–339.

Ram, U., 1999. The colonization perspective in Israeli sociology. *In*: I. Pappe, ed. *The Israel/Palestine question*. London: Routledge, 49–71.

Rempel, T., 1997. The significance of Israel's partial annexation of East Jerusalem. *Middle East Journal*, 51 (4), 520–534.

Reuchamps, M., Kavadias D., and Deschouwer, K., 2014. Drawing Belgium: using mental maps to measure territorial conflict. *Territory, Politics, Governance*, 2 (1), 30–51.

Saarinen, T.F., 1973. Student views of the world. *In*: R.M. Down and D. Stea, eds. *Image and environment*. Chicago, IL: Aldine, 148–161.

Shafir, G., 1999. Zionism and colonialism: a comparative approach. *In*: I. Pappe, ed. *The Israel/Palestine question*. London: Routledge, 72–85.

Sieffert, D., 2004. *Israël Palestine. Une passion française: la France dans le miroir du conflit Israélo-Palestinien*. Paris: La Découverte.

Smiley, S.L., 2013. Mental maps, segregation, and everyday life in Dar es Salaam, Tanzania. *Journal of Cultural Geography*, 30 (2), 215–244.

Smith, C.D., 2010. *Palestine and the Arab Israeli Conflict: a history with documents*. Boston, MA: Bedford/St. Martins.

Tolman, E., 1948. Cognitive maps in rats and men. *Psychological Review*, 55 (4), 189–208.

Tversky, B., 1993. Cognitive maps, cognitive collages and spatial mental models. *In*: A.U. Frank and I. Campari, eds. *Spatial information theory: a theoretical basis for GIS*. Berlin: Springer, 9–20.

UNRWA, 2007. The United Nations and Palestinian refugees [online]. Available from: https://www.unrwa.org/userfiles/2010011791015.pdf [Accessed 17 August 2017].

Waterman, S., 1980. Alternative images in an Israeli town. *Geoforum; Journal of Physical, Human, and Regional Geosciences*, 11, 277–287.

Weizman, E., 2012. *Hollow land: Israel's architecture of occupation*. London: Verso.

Yiftachel, O., 1999. Between nation and state: "fractured" regionalism among Palestinian-Arabs in Israel. *Political Geography*, 18 (3), 285–307.

Yvroux, C., 2012. *The Israeli-Palestinian conflict in representations*. PhD thesis in geography, Paul Valéry University, Montpellier.

Mental maps of global regions: identifying and characterizing "hard" and "soft" regions

Clarisse Didelon-Loiseau, Sophie de Ruffray and Nicolas Lambert

ABSTRACT

This article presents the quantitative synthesis of mental maps that identify different types of world regions. It is the result of a large-scale survey conducted in 18 countries, based on a sketch map approach. The number, shape, and extension of these vernacular world regions vary according to countries, cultures, and the personal styles of respondents who drew the maps. However, when we collectively analyze the regions identified by respondents, we observe that the figures of global regions are more or less recurrent. While the most commonly used division of the world is into "continents", we can identify "hard" and "soft" regions of the world. Whereas a "hard" region, such as Africa, can be recognized relatively unambiguously as a continent, "soft" regions may include numerous regional distinctions such as East Asia, Russia, South East Asia, and the Middle East. Our methodology involves defining a set of characteristics that discriminate between "hard" and "soft" regions (measuring spatial uncertainty and the relative vagueness of limits and fringes), then accounting for the correlation of these areas on the world map.

Introduction

Different disciplines in human and social sciences have analyzed "mental maps", i.e. the organization of spatial information in the human brain, although there are variations in the methods employed. Psychologists were among the first to focus on cognitive space, with the hypothesis that space is full of meanings and values. Such meanings are specific to each individual, but with collective studies, it is possible to determine more- or less-shared characteristics. Although spatial representations are essentially individual and molded by one's experience of space, exposure to different sources of information about places (classroom education, television, printed press, and the Internet, among others) also contributes to forming these representations. Individual spatial representations are, therefore, in part determined by community cultures and representations.

In geography, the investigation of mental maps has usually involved the production of maps that can be analyzed using quantitative methods and produced at different scales (Lynch 1960, Frémont 1976, Moles and Rohmer 1978, Cauvin 1997, Gould and White 1997). Most mental maps research has been conducted at city (or even neighborhood) level. At the national or supra-national level, some mental mapping research has focused on vernacular, or perceptual, regions, i.e. regions identified and named by ordinary people. Those terms are mainly used in the English-speaking geography and most of the analysis has been developed using United States vernacular regions (e.g. Zelinsky 1980, Lamme and Oldakowski 1982, Shortridge 1987, Colten 1997). Geographers such as Zelinsky (1980) and Shortridge (1987) have attempted to quantitatively delineate the degree to which a vernacular regional label is applied to particular areas, by using data sources such as business names and mapping exercises. Few studies, however, have focused on vernacular regions at the global level.

Since the eighteenth century, many regional divisions of the world have been proposed by social and natural scientists. Most of the initial divisions of the world would today be classified as formal regions, based on observable characteristics such as climate, population density, GDP per capita, or level of development (Poon 1997). Divisions of the world based on flows and networks, also known as functional regions, appear precisely when the world economy started to create more complex interconnections through globalized trade and migration (Chase-Dunn 1999).

Most relevant to a study of mental maps, however, is the category of vernacular or perceptual regions, regions that are seen by the general public as existing and sharing particular characteristics. In the analysis of world regions, continents hold a particular position as the most frequently observed regionalization of the world. The exact number, shape, and land area of such perceived regions vary depending on the country under study and its prevailing academic traditions. Continents are long-standing figures in the division of global space, inherited from the worldview held by the ancient Greeks, who organized the world into in three sections (Europe, Asia, and Africa) around a central area (the Aegean Sea). This categorization, and the boundaries it involved, was soon disputed, most notably by Herodotus, but it continued to serve as the foundation of a Christian worldview which divided the world between the three sons of Noah.

With the emergence of modern sciences and their passion for classification, continents were "naturalized" (despite their lack of grounding in objective observations) and became one of the most common models of geographic knowledge, particularly in the classroom. The widespread use of continents in cartographic representations has established their prominent position in contemporary representations of the world: they are one of the principal forms of meta-geography (Lewis and Wigen 1997), the set of spatial structures through which individuals order their knowledge of the world. Continents, therefore, fall within the production of regionalization that we use to organize incomplete and often imprecise world knowledge and to manage our limited abilities of memorization (Ellard 2009) through the process of categorization, a universal characteristic of human thought (Montello 2003).

Mental maps research at global level was initiated by T.F. Saarinen with the aim of demonstrating the variety of points of view of the world or of testing the level of geographic knowledge on some samples of the world's population (Saarinen and MacCabe 1995, Saarinen 1998). These analyses were most often qualitative and focused on the centering of maps according to the place of residence of the people surveyed (Saarinen 1987) or on the number and size of the countries drawn. Pinheiro (1998) sought to explain the frequency of representations of location via statistical approaches based on the gravity model.

Vernacular regions are defined by those who perceive them through a broad range of criteria, such as visual appearance, specific activities, ethnic composition, and many other characteristics not strictly related to spatial cognition (Kuipers 1978). They are created by using and mixing various sources of information (Battersby and Montello 2009; Friedman 2009). The locations of these regions cannot necessarily be ascertained with precision, precisely because they are not rooted in any sort of objective criteria. Indeed, one of the characteristics of vernacular regions is their fuzziness and impreciseness. A French geographer, Grataloup (2009), in *L'Invention des continents* (a book that has not been translated into English) has suggested the existence of "hard" and "soft" global regions which differ in the way they are perceived. In the former, one has no hesitation about which global region it is (in Africa, in Europe). In the latter, numerous instances of a sense of continental membership may be mentioned (Maghreb, North Africa, and Africa). The book provides a map which represents "the degrees of ambiguity of continental membership to global regions". He empirically constructed and differentiated five types of spaces according to the stability in their representations (hard continents, strongly autonomous regions, intermediary regions, shared regions, and poorly situated peripheries). These concepts, by focusing on the degree of agreement over regional labels affixed to particular locations and the level of agreement as to their boundaries, can add a degree of precision to our discussions of vernacular regions.

This article is an attempt to incorporate the concepts of hard and soft regions into our understanding of global vernacular regionalization. We examine the mental regions of students from 18 countries. We then examine these representations to characterize hard and soft regions and to determine the factors behind the formation of such cognitive structures. The analysis presented here is different from most of studies on mental maps at world level because it does not focus on regions within particular countries, but on global regions. The methodology used is also different, particularly in our use of a GIS that allowed us to quantitatively synthesize thousands of individual maps, in order to highlight the main shared structure in representations of global regions.

Sources and methods

The EuroBroadMap survey

The analyses presented here are based on the results of a survey conducted in a European research project EuroBroadMap: Visions of Europe in the World,[1] whose goal was to identify the spatial perceptions of Europe and the world. The major goals of the project were to explore the ways that Europe is represented in people's minds, in political discourse, and in theories of international relations, and to analyze how its representations vary according to the country surveyed and how it is embedded in functional regions and networks (Didelon and Grasland 2014).

We used mental mapping methods for the collection and analysis of data. The questionnaire used "interpretative" type mental maps where the respondent is asked to identify a spatial phenomenon (Didelon *et al.* 2011). We provided a blank base map representing the outline of spatial units (the countries) as polar projection (centered on the North without favoring any of the locations in the survey[2]). Respondents were then asked to delineate the regions of the world and name them. No specific instructions were given regarding the possibility of drawing regions which overlap, so this was implicitly permitted. However, in most cases, the maps produced presented very few overlaps. This base map then made it possible to arrange the individual responses collectively and allowed for an aggregate analysis of the responses (Kitchin and Fotheringham 1997). The question was formulated in this way: "On the following map, draw your own divisions of the world (maximum of 15). Which names would you give to each of the areas?" This question made it possible to understand the makeup of geographic regions which are both groups of places judged to have common characteristics and products of an interpretation of the world which can be analyzed. The aim was to comprehend the regionalization of the world that exists in the respondents' minds. How many world regions were drawn? What were their characteristics (name, position, size)? Where were their boundaries located? And were they concentrated or fuzzy; in other words, how much agreement was there about where the borders of the region are? This allowed us to determine the existence of "hard" and "soft" regions in the minds of our respondents.

The survey was conducted internationally in 43 towns and 18 countries[3] (Figure 1) of almost 9500 third-year undergraduate students from six disciplinary fields (art, health, engineering, geography, economics, and political science) who responded to the questionnaire provided in their respective languages. Students at this level have generally acquired a disciplinary identity that can influence their representation of the world. They are also in the process of choosing their specialization and beginning to develop a relatively clear idea about their future profession, the sectors in which they would like to work, and their professional goals certainly had an influence on their view of the world.

The countries surveyed were chosen in order to represent a broad range of relational experiences with respect to the European Union (EU), with different levels of economic development, and a wide diversity of geopolitical positions Firstly, there were countries that were members of the EU from the start (France and Belgium), then countries that symbolized enlargement of the EU to the south (Portugal, 1986) and to the north (Sweden, 1995), and finally countries that became members only during the last rounds of enlargement in 2005 (Malta, Hungary) and 2007 (Romania). Other sample locations are located on the fringes of the EU and are officially registered candidates (Turkey), or have been more or less beneficiaries of policies in neighboring countries (Moldova). Lastly, some countries share a historical experience

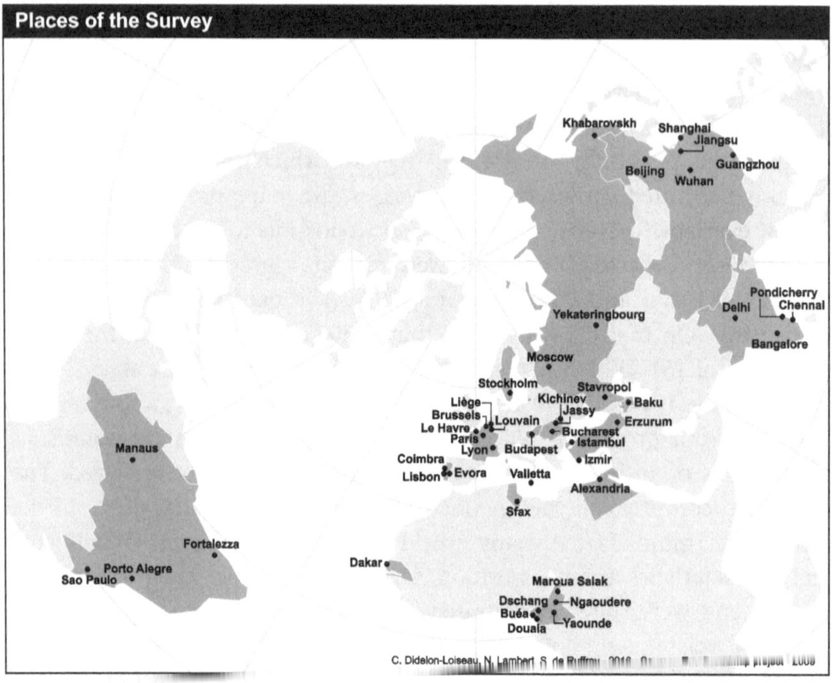

Figure 1. Places of the survey.

linked to different phases of colonization (Brazil, Cameroon, Egypt, India, Senegal, Tunisia), while others have different, and more independent, relationships with Europe (Azerbaijan, China, Russia). The towns selected are both capital cities (because all fields of studies are present) and small- or medium-sized towns situated nearby (Le Havre, Lyon and Paris; Evora, Coimbra and Lisbon). The towns also have different characteristics as regards their openness to the outside world (particularly concerning the selected in China and Brazil).

Whereas we developed the questionnaires *in situ*, the scan of all maps was done by one team with one unique device in order to minimize the drawing distortions. Digitization of regions was done semi-automatically on the ArcGIS platform in order to build a georeferenced database. Each of the regions mapped out by a student is represented in the database by a polygon (the region), the name ascribed to it, and also some attributes of the student who created it, in particular, the country and town of the survey. Students wrote the names of the regions in their own languages (we conducted the survey in nine languages). The national researchers involved in the project then translated the names into English. Any questionnaire containing an ambiguous outline (for example a region not delineated by a closed line) was removed: no questionnaire was partially digitalized. In total, 4481 questionnaires (47% of the questionnaires collected) were usable for the analysis. The database constructed contains almost 30,200 named polygons (an average of 6.7 regions per questionnaire), which we then superimposed on a world map.

Types of regions obtained

Several main patterns in representations of world regions can be seen in the survey results. The most commonly used representation of the world was the division into continents. Most of the students used "geographic" terms to describe their representation of the world and these terms refer mainly to the continents' traditional names. One fact supporting this finding is that 41% of students drew 5–7 world regions that corresponded to the division of continents in the different national academic traditions according to scholarly books gathered and analyzed by our project partners in each country. Within the same country, regions can also be taught differently depending on the discipline (e.g. geology, geography).

Furthermore, the definition of their boundaries raises questions. Some boundaries can more easily and more consensually be identified than others. This is especially true of Africa, almost totally surrounded by maritime spaces (with the exception of the narrow Suez isthmus. The demarcation of Europe on its eastern side, meanwhile, is much more ambiguous. However, despite its fuzzy boundaries, Europe seems quite well rooted in people's minds, as stressed by Lewis and Wigen (1997).

The second-most common source of regional identification was the division of the world according to economic and development characteristics, or by geopolitical identity. It was observed in about 20% of the questionnaires. For example, the names given to the regions in one of the questionnaires were "very developed", "developed", "emerging", "developing", "poor", and "very poor". As pointed out by Lewis and Wigen (1997), this is also a commonly used meta-geography, shaping how we perceive the world. Such regionalizations are familiar to students, and they are widely disseminated by United Nations institutions. In such cases, the world is no longer divided into neutral, geographically defined regions, but rather as a collection of geopolitical oppositions or development hierarchy levels. Students often provided their personal interpretation of the difference between regions. For example, the following names were given: "natural resource rich developing third world countries dependent on first world countries" or "has shown no development dependent on the rest of the world exploited countries". Sometimes they express value judgments that could be patently politically incorrect and rather aggressive and racist like the following: "countries go against & resist blood sucker countries", "North America where at least as cold as its weather cold people live" or "black Africa corruption". Among those hierarchical regionalizations some questionnaires made only two regions and used the coupled categories "north/south" or "west/east", "developed/undeveloped", "rich/poor", "exploited/exploiters". Such was the case for 4.2% of the questionnaires.

Some students have developed an even more personal representation of the world: they have a specific poetic vision of the world, inventing poetic shapes and names to portray the regions (for example: "sonthesiya" or "lokoria". They also drew some regions and built a specific and rather personal discourse about the places identified; they described their feelings about the places with descriptions such as "I dream to be these places" or "I want to visit". The most "personal" ones are the regionalization maps where "I" and "my" are used in the descriptions of world regions. Here the students develop a fairly personal vision of the world, positioning themselves and appropriating some places as "my home", "my country", and developing projects "where I would like to travel". Their practical experience of the world or will to experience it (or not) is often expressed in these kinds of world representations.

Methods of identifying hard and soft regions

The first aim of this article is to define which characteristics may be observed and quantified in order to identify "hard" and "soft" regions. Their analysis and representation involve theoretical and methodological reflections on the definitions, membership, and boundaries of these geographic areas. These "hard" and "soft" regions belong to the category of subjective spaces, the mental representations of geographic space. They are thus the result of a twofold process combining different perceptions of geographic space and construction of knowledge from individual and collective patterns (Pocock 1976, Cauvin 1997). Regions are represented, to a greater or lesser extent, in relation to one another, and collectively define a spatial organization of greater or lesser complexity.

Subjective spaces are mental regionalizations of global space and are also characterized by uncertainty. They are linked to the respondents' capacities to deal with the survey according to their education, perception, personality, culture, background, and so on. Methods and concepts founded on fuzzy logic are particularly relevant for grasping the mental representation of mental regions that are characterized by a high level of vagueness and uncertainty (Rolland-May 1986, 1987, de Ruffray 2007, Didelon *et al.* 2011). They make it possible to identify areas of confusion and uncertainty with regard to continental membership (Burrough and Frank 1996). Each region drawn by a group of individuals represents a fuzzy geographical space. For each component of this geographical space, i.e. a cell of the grid, a membership function has been defined. We characterized a region as part of a "hard" region when the number of possible belonging to a region given to it by respondents was unique or very small. Conversely, a location is classified as part of a "soft" region is when the number of possible areas of membership increases. A collection of locations constitutes a "hard" region when the boundaries drawn between it and neighboring regions are concentrated – that is to say, when there is a high degree of agreement about where the border of the region is, and there is a sharp discontinuity between the regional identifications assigned to it and neighboring regions. Conversely, a "soft" region has boundaries that are characterized by slowly decreasing numbers of respondents who assigned an area to that region.

We identified regions based on the names associated with the outlines drawn by each student. This approach works only if a significant number of students have used the same name. If there are numerous expressions used by students to name regions, we focused on the names used most consistently to define certain spaces. In the analysis reported here, we retained only the names of regions named more than 100 times, which in this study amounted to 32 different regional labels. We grouped names referring to strongly overlapping geographic space are grouped together (e.g. South America and Latin America are combined under the term SouthLatin America and Australia and Oceania under the name AustrOceania). While it is true that in many areas of the world, these terms have distinct meanings, because of the high degree of confusion between these regional labels in some regions where our study took place, we have combined them here. Additionally, because all of the responses were translated from the students' original replies, we did not know what was actually written by students; translators in some cases may have conflated the two terms or used them interchangeably.

The word "poverty", used by some respondents to name some regions, but a term not typically used to refer to a specific geographic space, was excluded from this analysis. We also excluded "Antarctic", a place which did not appear on the map used for the survey but was often confused with Oceania due to the polar projection employed. In the end, we retained 30 names of regions for further analysis.

In order to examine this spatial information in a quantitative and aggregate manner, we defined a regular geometric grid (in the polar projection defined in the questionnaire) composed of squares approximately 150 km across, for a total of 66,885 cells. For each drawing, we calculated the regional identity (from among the 30 regions retained in the analysis) assigned to each of the grid's cells; if a cell was split between two regions, we assigned it the region that encompasses the larger part of the cell. For every regional name, we created a database showing the number of respondents who assigned each cell to that region. We then analyzed the data, and created a three-dimensional visualization of the data, representing the strength of association with a particular name as a topographic surface; we then were able to employ various forms of topographic analysis, including slope, drop, gradient, and plateau analysis. The height of the surface represented the number of respondents who assigned that regional name to different areas, and the slope, drop, gradient, and plateau analysis could determine the degree of ambiguity and disagreement about where the borders of the region lie.

This first step allows for identification of the great spatial structures which correspond to the regions that dominate in mental representations of global space. A map of the data representing the most-used regional name gave us an initial simple organization of world regions. On this global regionalization, six continental structures appear: North America, SouthLatin America, AustrOceania, Asia, Africa, and Europe. If this division refers to an obvious worldview – the division into continents – a second level of analysis provided us with a greater degree of understanding. By mapping the second-most common regional name given to each cell, we were able to see other subregions emerge such as the Middle East and Russia.

We defined the degree of "hardness" or "softness" of regions identified by the students according to both citation intensity and also the shape of the transition to other regional terms found at its boundaries. The citation intensity is used as a first step to identify two principal types of global region: on the one hand, "hard regions", which are frequently named and drawn by the survey's respondents; on the other hand, "soft regions", which are cited at lower intensity. The hard regions were all cited by more than 1500 students whereas the soft regions were cited by a maximum of 800 students. The difference in value shows how extensively certain global-scale regional structures dominate in mental representations, leaving little room for other types of regions in people's mental maps.

We further analyzed the shape of the regional perceptual map by drawing a series of transects of the "surface". Here we represent the citation intensity per square (Figure 1). This analysis, borrowed from topographical cross-sectioning in physical geography, enables the "relief" of the variations in the membership areas to be visualized. Both map (Figure 3) and transect (Figure 2) are employed together to identify and characterize the regions that structure the mental representation on a global scale. The map was used to identify the spatial configuration of the regions drawn by students that we identify as hard regions. Transects were used to characterize both the intensity of the quotation of the region name and the shape of the slope (discontinuity or gradient).

Hard and soft regions: analysis and discussion

Hard regions

The dominant regional structures were unambiguously continental names. However, even among the continents, there were broad differences in citation

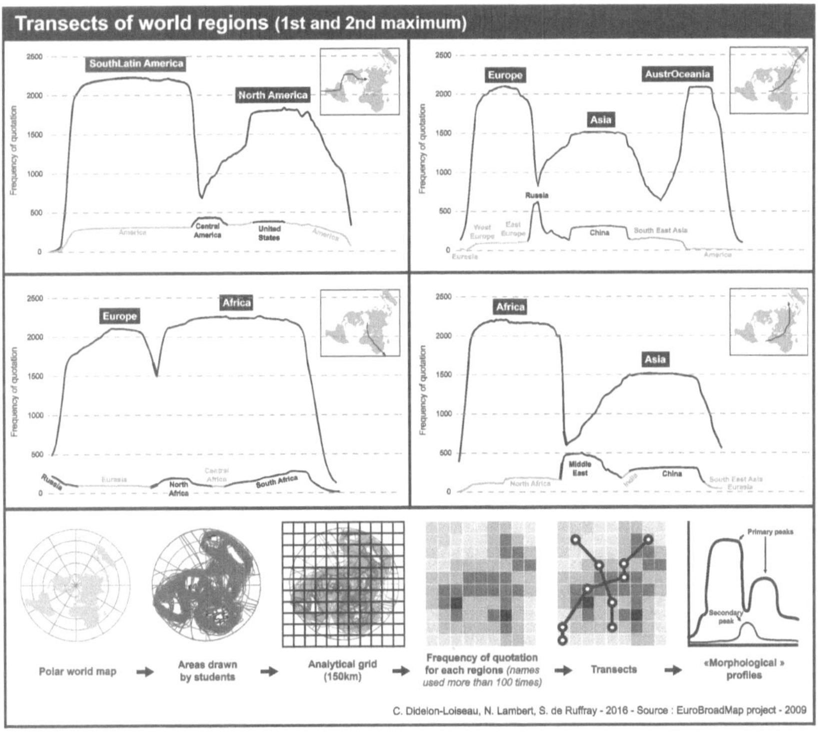

Figure 2. Transects of world regions.

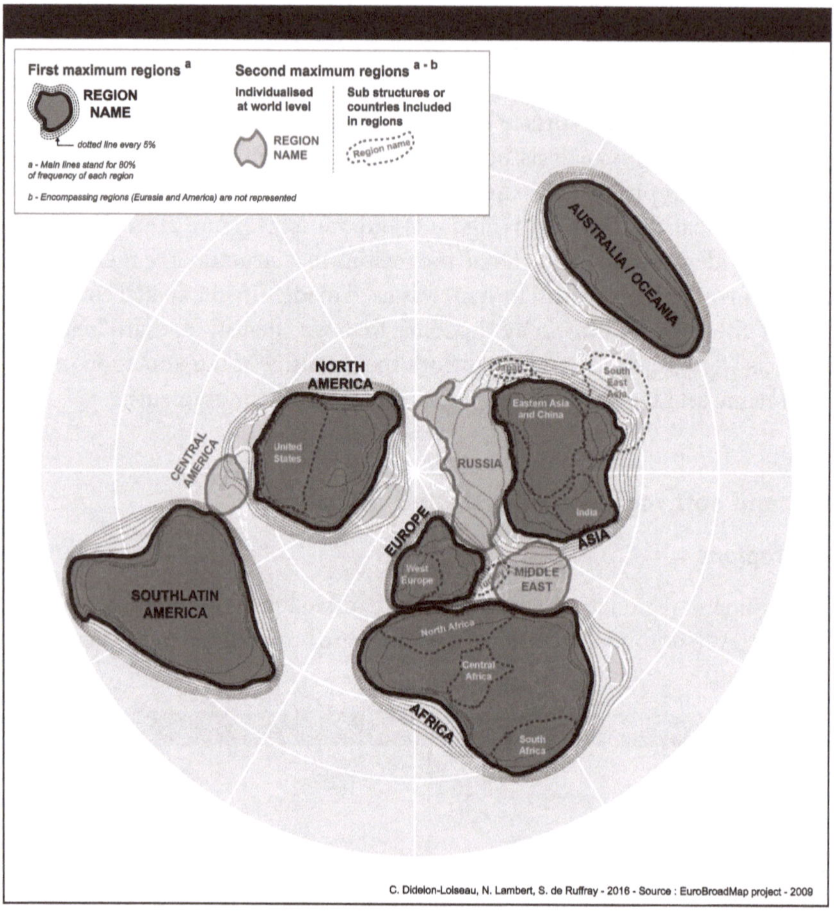

Figure 3. Quantitative synthesis of world regions.

levels: Africa was cited 2387 times, SouthLatin America 2389 times, AustOceania 2323 times, Europe 2293 times, NorthAmerica 1950 times, and Asia 1722 times. Visually, on transects, these regions take the form of high plateaus with steep inclines (except for Asia and NorthAmerica). Such regions, with large variations in altitude over a short distance, indicate the existence of a strong degree of consensus on their boundaries. Among the most cited regions, only Asia and North America depict a less-defined shape with gentler slopes: this indicates a slower decrease in membership values as you move away from the core area of the region. Their altitudes, furthermore, are not as high, which indicates that they are the least recognized of the hard regions.

The map (Figure 3), which gives a quantitative synthesis of the regions identified, shows that there are variations in the land area of these *hard regions.* On this map, the region identified by 80% of students as Africa displays the largest surface area (since Russia was often excluded from Asia, and treated either as part of Europe, or as a separate region), and Europe the smallest. The size of the plateaus that we previously identified depends on the land area of the continents drawn on the map but also on the orientation of the transect: in transect 3, Africa is depicted from north to south, while in transect 4, it is depicted from east to west. These mental regions, therefore, by and large, follow the same forms and land areas of continents in the way they have usually been identified by classroom traditions in the countries surveyed.

If one takes into consideration, the respondents who named and drew each region, we can identify some significant patterns which make it possible to characterize the regions' boundaries. Regions are most strongly delimited along maritime edges, most notably the seas and oceans along the shores of Africa and South America. Most of the boundaries are drawn at a relatively small distance from the coasts and give rise to clear discontinuities which are clearly identifiable on the transects. The boundary gives rise to a relatively gentler gradient when an "object", most often a large island, casts doubt on the boundary's location: this is true of Madagascar with respect to Africa, of Greenland for North America, and of Papua (one part of which is part of Indonesia and the other forms an independent state: Papua New Guinea) with respect to AustrOceania. This predilection for maritime spaces is also conveyed by a great concentration of boundaries between two regions in straits like the Strait of Gibraltar and the Bosporus and Dardanelles straits.

There are two general patterns we found when perceptual boundaries cross landmasses. The students' surveys either minimize the landmass by choosing to draw the boundaries at isthmi, like Panama; or else they make them carefully follow national borders: this is the case with respect to the border between Panama and Colombia, but also over even longer distances like the border that separates the United States and Mexico or even the Russian Federation on the one hand and Finland, Estonia, Lithuania, Belarus and Ukraine on the other. These two graphical solutions express the aversion the respondents show towards drawing boundaries which would chop a country in half, a case which happens only rarely.

Finally, the structure of the boundaries appears to vary according to the nature of the space crossed (terrestrial or maritime) and according to the proximity of other hard regions. Consequently, the membership values for the regions of "Africa" and "Europe" are still very high (around 1500) when one passes from one region to another: their boundaries cross a relatively narrow sea, the Mediterranean, and the two regions are relatively near to one another. With respect to land boundaries, the distinctions are generally softer, with vast areas of uncertainty appearing. These spaces can sometimes be marked by a strong asymmetry in shape and gradients. For example, the terrestrial transition between SouthLatin America on the one hand, and

North America on the other, is marked by a clear discontinuity at the edge of South/Latin America, while North America has a much gentler gradient depicting its boundary, indicating a greater degree of consensus about the edges of the former when compared with the latter, whose perceived bounds sometimes included all or part of Mexico as well as the countries situated between Mexico and Colombia. The same situation is seen with respect to the boundary between Europe (which has a clear discontinuity at its border) on the one hand and Asia (which has a more gradual gradient) on the other. North America and Asia are the regions we identified as the "softest" of the "hard" regions, due to the lower frequency of citations and the shape of gradients that characterize their boundaries.

The gradients identified on the transects reveal that they are situated in specific spaces, in global regions that are not among the hard regions. These areas of transition between continents are particularly interesting to analyze. They usually represent an area of hesitation or disagreement when allocating locations and this is expressed on the students' drawings by intersections or overlapping of the regions drawn. These spaces lacking strong membership reveal another level of structure, the "soft" regions. The fuzzy limits, therefore, leave room for other regional structures (Brunet 2001).

Soft regions

The regions that correspond to the second maximum are characterized by relatively low membership values, ranging from 101 mentions (Central Africa) to 793 (Russia), with only Russia and the Middle East (553) exceeding 500 citations. Even if the strong regions clearly dominate mental representations, soft regions also appear frequently on the map. It is impossible to determine a threshold for their appearance. Their characteristics are explained more by citation intensity and their location, notably with respect to hard regions.

The soft regions that show the greatest intensity are located in spaces of intersection between two continents, but above all in spaces where there are asymmetric discontinuities – in other words, spaces marked by a great discontinuity on one side and a gentler gradient on the other. These regions are located on the map in the "gaps" left by strong regions. This is the case for three of the most-cited soft regions, but also for South East Asia, which is relatively less frequently cited:

- Russia is positioned between Europe, whose borders have a high degree of agreement, and Asia, whose membership gradient rises more slowly from west to east. Russia is a rather confined area, given the orientation of the transect, but is the soft region showing the strongest membership values. This can be explained by the fact that Russia is a state bounded by internationally recognized borders which makes identification simpler. Its great expanse and socio-cultural features make it a distinct region between the Europe and Asia. Finally, the prevalence of marked nationalism in Russia explains why many of the respondents from Russia chose to single out their country on a world map (Didelon-Loiseau and Richard 2015).

- The Middle East is positioned at the intersection of three regions: Africa, Europe, and Asia. The transect illustrates the transition between Africa, whose membership values drop sharply and Asia, whose membership values increase slowly. Situated on the fringe of the Asian region on the transect, this region nevertheless appears in a space where the membership values for Asia do not exceed 20%. Its shape on the transect is also asymmetrical with a marked discontinuity on the African side and a gradual reduction in membership to the region as membership to Asia increases.
- The same observation can be made for Central America, located in the interstice between South and Latin America on the one hand and North America on the other: it extends principally in an area of low membership to North America whereas there is a marked discontinuity of membership values to South America.
- South Asia is the weakest of this type of region in terms of frequency, being located at the intersection of two regions marked by the gradients of Asia and AustrOceania. The map shows that, while being partially integrated into Asia, it extends also into an "empty" space and does not correspond to values lower than 20% of membership to AustrOceania. This can probably be explained by the physical configuration of the space: a vast maritime area marked by the presence of several islands which is likely to complicate identification of a region since large landmasses play an important role in identifying regions, as we have explained.

The second type of soft region is mainly made up of sub-regions included within hard regions. They are often situated in positions on the fringes or edges of the hard regions, as is the case for North Africa, South Africa, Western Europe, India, China, and the United States of America, but without actually being at an intersection with another region. Here, it is probably the knowledge of internal divisions within spaces that explains the appearance of these sub-regions, which are often marked by a strong identity compared with the rest of the space in which they are located and which make it possible to individualize them, as is the case for the United States as the most powerful country in the world or India with its specific culture. The knowledge of internal structures of certain continents or global regions is nevertheless not widespread: this is particularly true for Africa and is probably explained by an ignorance of its internal diversity. Consequently, identifying Central Africa in the course of our survey can most likely be explained by the presence of a sub-sample in Cameroon, situated precisely in this region of Africa.

The region of "Japan" is a particular case (not described here by transects): it is not situated at the intersection of two regions but at the edge of Asia. On a global scale, it appears as if it was singled out and the membership values to Asia are relatively weak. As a result, being under the threshold of 80%, it is not possible, according to this criterion, to consider it as totally "included" in Asia. This particular positioning can be explained by several factors: Japan is a country distinguishing itself from the rest of Asia, notably in terms of wealth and standard of socio-economic development; it is also a country marked by a strong cultural identity and finally, historically, its relations with its neighbors have been rather conflictual. The result of all these elements is to consider it as a distinct space in the Asian region.

The last major type of soft region which can be identified is made up of vast regions stretching over a substantial proportion of global space and includes several "hard" regions seen above: America and Eurasia include, respectively, North America and South and Latin America on the one hand and Europe (see transect) and Asia on the other. Such mega-regions were little cited and appear as a top choice only in the interstices left by the hard regions and the different types of soft regions previously mentioned.

Discussion

This analysis enabled us to identify and estimate the strength of different world regions in mental representations of students from different countries. One main finding of this analysis is that, despite the strong differences in countries of origins and disciplinary fields, the students' representations share some significant common features. Indeed, the analysis of the mental maps on a global scale collected in the framework of the EuroBroadMap project confirms the predominance of global representations structured by continents regardless of where the survey is carried out. This means that, from Brazil to China and from Sweden to Cameroon, continents are a common basis for regional perceptions and descriptions of the world. Such continental-scale mental mappings of the world have been well analyzed as has its role in the distortions and simplifications of global understanding (Lewis and Wigen 1997, Grataloup 2009). However, this relatively well-shared mental representation of the world is, for many people, seen as naturalized and "free of ideology".

Such perceptions likely are fostered by the common occurrence of international organizations, such as U.N. agencies, large NGOs, and multinational companies (Didelon, 2010), but also at a national level (through diplomatic services, travel agencies, and other entities) that are founded, if not on the boundaries of continents, then at least on their name. For example, the

Food & Agricultural Organization proposes a division of the world mixing development levels and geography, based on continent names. As a consequence, the developed zone is divided into "industrialized developed countries" and "transition developed countries", while developing countries are classified according to their regional location (South America & the Caribbean, South Asia, etc.) (Didelon 2005). Such regional constructs are also given a firmer place in the mental maps of people by their use in the names of areas of economic integration such as the EU, the North American Free Trade Agreement, and the Association of Southeast Asian Nations. Therefore, the figures of continents continue to be the central organizing regional construct in a globally shared representation of the world.

It is also possible, as we have shown, to identify both hard and soft regions in representations. This allows for observation of a global space in which Africa is the "hardest" continent, whatever the reasons for this hardness might be (affirmation of a perceived homogeneity or ignorance of an actual heterogeneity) whereas Asia is the "softer" continent among the "hard" regions, characterized by lower citation intensity, the absence of a strong discontinuity of its boundaries, and particularly by the existence of rather "strong" secondary regional structures on its edges and numerous sub-divisions within it. The maps and transects give a clear view of the location of "soft regions" and make it possible to explain their presence through the existence of sub-systems (Brunet 2001) on the fringes of "hard" regions or within the "softest regions".

The existence of these hard and soft regions raises many theoretical questions for the analysis of mental representations of the world which may have consequences in terms of mutual understanding between the people of world and regional affairs. Does the existence of hard regions (like Africa) really result from the agreement of the respondents on the definition of some locations, or does it stem from the lack of knowledge about the internal diversity of Africa? The first would have positive consequences in mutual understanding, while the second raises troubling questions about global levels of ignorance of Africa. Similarly, does the existence of soft regions in students' mental maps (like the Middle East) result from the disagreement of students on the delineations of hard regions, or from the acknowledgment of unique cultural, economic, or geopolitical issues in this area? At this stage of our research, it is not possible to determine explicitly the causes of regions' level of hardness or softness.

This study only examines the most-used regional terms – those used by more than 100 students. Other, less-used names drew on a broader variety of criteria, including climate, civilization, culture, history, geopolitics, economics, level of development, and even on individual appreciations. Each individual apprehends the world from a mix of thematic composites and more or less well-known regions. As stated by Polonsky and Novotny (2010), that means that the vernacular regional structures of the world derive mainly from education and collective knowledge spread out by media. They can include "continents" but also "civilizations" or "development levels". However, those "academic" regions are more or less known and people also use their own experience and perceptions to divide the world in regions.

Lastly, another finding of this research is the place of maritime spaces in global mental representations. Firstly, maritime spaces are used by students mostly to draw the boundaries between global regions. They seem reluctant to draw boundaries that cross landmass areas. Secondly, if boundaries are located in maritime spaces, only a very small proportion of seas or oceans are included in global regions. Indeed, for students, the world seems to be constituted solely of landmasses: most boundaries drawn are very close to coastlines. This absence of the sea in global regionalization or its use to locate boundaries shows the minor importance given to the sea in representations of the world. Indeed, maritime spaces are never included in global regionalization produced in school books or in publications of international organization. Only the landmasses where human populations and activities are located are taken into account in world representation, even if seas cover the major part of the planet (71%). This absence of the sea in global representations is a cultural issue because it stresses how restricted membership to the maritime space is despite playing a major role in both globalization (maritime routes, fishing activities) and global environmental issues (global warming, microplastic pollution).

Conclusion

This paper analyzed mental representations of global regions of a large number of students from 18 countries. This empirical analysis provides a complementary perspective on global regions and continents to the theoretical academic literature. Although the use of continents as meaningful divisions of the world has been largely rejected by geographers, its terminology continues to be widely used by students to identify global regions. Therefore, continents constitute a fairly significant shared representation of the world that allows for mutual understanding. It also makes it possible to delve deeper into the classic analysis of world mental maps by using a method that provides a quantitative and mappable synthesis of numerous spatial representations. By testing concepts found in French geographic literature (hard and soft regions), this article also contributes to methodological advances in the analysis and depiction of mental representations of a complex space and especially in visualizing locations that are included in multiple regional conceptions. This article identifies vernacular regions of the world following mental representations of global regions in the world level by using quantitative methods.

Further stages of our research will explore variation in representations of global regions from country to country (see Didelon-Loiseau and Richard 2015). We also plan to increase our sample of countries in order to check some of our hypotheses related to the explanation of variations in hardness and softness, mainly in Japan, USA, and Australia but also in Africa and Central Asian countries (a project launched in September 2017 will allow us to collect surveys in Kazakhstan).

Notes

1. The EuroBroadMap project was funded by the European Commission in the framework of the Seventh Framework Program (www.eurobroadmap.eu).
2. We chose a polar projection because it was the only one that was unfamiliar to all the students surveyed. This was tested and established during the test phase of the survey.
3. No participants from the North American or Pacific regions were included, due to the lack of interested potential scientific partners in those regions. Certainly, their inclusion would have affected perceptions of North, Latin, and South America.

Disclosure statement

No potential conflict of interest was reported by the authors.

Funding

The research leading to these results has received funding from the European Community's Seventh Framework Programme (FP7/2007-2013) under grant agreement n° 225260 (www.eurobroadmap.eu).

References

Battersby, S. and Montello, D., 2009. Area estimation of world regions and the projection of the global-scale cognitive map. *Annals of the Association of American Geographers*, 99 (2), 273–291.

Brunet, R., 2001. *Le déchiffrement du monde, théorie et pratique de la géographie*. Paris: Belin, coll. Mappemonde.

Burrough, P.A. and Frank, A.U., 1996. *Geographic objects with indeterminate boundaries*. London: Taylor & Francis, GISDATA series.

Colten, C.E., 1997. The land of Lincoln: genesis of a vernacular region. *Journal of Cultural Geography*, 16 (2), 55–75.

Cauvin, C., 1997. Proposition pour une approche de la cognition spatiale intraurbaine, Cybergeo. *European Journal of Geography*, 72. Available from: http://journals.openedition.org/cybergeo/.

Chase-Dun, C., 1999. Globalization: a world-systems perspective. *Journal of World Research System*, 5 (2), 165–185.

Didelon, C., 2005. Mental maps: Firms, Countries, International Organisations and NGOs' World regionalisation, in Grasland et al., *Second intermediary report of the ESPON 3.4.1. Project, "Europe in the world"*, volume 1, 89–120. Available from: https://www.espon.eu/programme/projects/espon-2006/coordinating-cross-thematic-projects/europe-world.

Didelon, C., 2010. Quand les multinationales divisent le Monde. Représentations et pratiques du territoire-Monde. *Revue Géographique de l'Est* [online], 50 (1–2). Available from: http://journals.openedition.org/rge/.

Didelon, C. and Grasland, C., 2014. Internal and external perceptions of EU in the world through space and time. *In*: N. Chaban and M. Holland, eds., *Communicating Europe in the times of crisis: external perceptions of the European union*. Baden-Baden: Palgrave-McMillan, 65–96.

Didelon, C. *et al.*, 2011. A world of interstices: a fuzzy logic approach to the analysis of interpretative maps. *The Cartographic Journal*, 48 (2), 100–107.

Didelon-Loiseau, C. and Richard, Y., 2015. Les étudiants russes sont-ils eurasistes? *Belgeo*, 4. doi:10.4000/belgeo.17624.

Ellard, C., 2009. *You are here: why we can find our way to the moon, but get lost in the mall*. New York: Doubleday.

Frémont, A., 1976. *La région, espace vécu*. Paris: Armand Colin.

Friedman, A., 2009. The role of categories and spatial cuing in global-scale location estimates. *Journal of Experimental Psychology: Learning, Memory, and Cognition*, 35, 94–112.

Gould, P. and White, R., 1997. *Mental maps*. 2nd ed. London: Routledge.

Grataloup, C., 2009. *L'invention des continents*. Paris: Larousse, coll. Terre et Nature.

Kitchin, R., Fotheringham, M.A.S., 1997. Aggregation issues in cognitive mapping, *The Professional Geographer*, 49 (3), 269–280.

Kuipers, B.J., 1978. Modeling spatial knowledge. *Cognitive Science*, 2, 129–153.

Lamme III, A.J. and Oldakowski, R.K., 1982. Vernacular areas in Florida. *Southeastern Geographer*, 22 (2), 99–109.

Lewis, M.W. and Wigen, K.E., 1997. *The myth of continents: a critique of metageography*. Berkeley: University of California Press.

Lynch, K., 1960. *The image of the city*. Cambridge, MA: MIT Press.

Moles, A. and Rohmer, E., 1978. *Psychologie de l'espace*. Paris: Casterman, collection "synthèses contemporaines.".

Montello, D.R., 2003. Regions in geography: process and content. *In*: M. Duckham, M.F. Goodchild, M.F. Worboys, eds. *Foundations of geographic information science*. New York: Taylor & Francis, 173–189.

Pinheiro, J., 1998. Determinants of cognitive maps of the world as expressed in sketch maps. *Journal of Environmental Psychology*, 18, 321–339.

Pocock, D.C.D., 1976. Some characteristics of mental maps: an empirical study. *Transactions of the Institute of British Geographers*, 1 (4), 493–512.

Polonsky, F. and Novotny, J., 2010. Cognitive mapping of major world regions among Czech geography students. *Journal of Maps*, 6, 311–318.

Poon, J.P.H., 1997. The cosmopolitanization of trade regions: global trends and implications, 1965–1990. *Economic Geography*, 73 (4), 390–404. Available from: http://www3.interscience.wiley.com/cgi-bin/fulltext/121592228/PDFSTART.

Rolland-May, C., 1986. Méthode de régionalisation d'espaces imprécis et incertains. *Revue de géographie de l'est*, 3–4, 213–242.

Rolland-May, C., 1987. La théorie des ensembles flous et son intérêt en géographie. *Espace Géographique*, 16 (1), 42–50.

de Ruffray, S., 2007. *L'imprécision et l'incertitude en géographie. L'apport de la logique floue aux problématiques de régionalisation*. Mémoire d'habilitation à diriger des recherches, université Paris 7.

Saarinen, T.F., 1987. *Centering of mental maps of the world. Discussion paper*. Tucson, AZ: Department of Geography and Regional Development.

Saarinen, T.F., 1998. Centering of mental maps of the world. *National Geographic Research*, 4 (1), 112–127.

Saarinen, T.F. and MacCabe, C., 1995. World patterns of geographic literacy based on sketch map quality. *The Professional Geographer*, 47, 196–204.

Shortridge, J.R., 1987. Changing usage of four American regional labels. *Annals of the Association of American Geographers*, 77 (3), 325–336.

Zelinsky, W., 1980. North America's vernacular regions. *Annals of the Association of American Geographers*, 70 (1), 1–16.

OPEN ACCESS

Changing mental maps of the Baltic Sea and Mediterranean regions

Janne Holmén

ABSTRACT

Little empirical research has considered the way in which macro-regions are perceived outside academic and political circles. Such studies alone can determine what regional narratives mean for the wider public, and the extent to which they coincide with region-building images produced by elites. This article examines the mental maps of high school seniors in 10 cities in the Baltic Sea and Mediterranean regions, focusing upon their perception and knowledge of other countries in those areas. Despite efforts at region building since the Cold War, the two regions remain divided on mental maps. Students have little knowledge of countries across the sea from their own, although such knowledge is generally greater among those from coastal (and particularly island) locations. A comparison with maps constructed by Gould in 1966 reveals that the perception of countries within one's own region among Italian and Swedish students has become more negative over the last 50 years.

Introduction

In 1975, Yi-Fu Tuan claimed that, despite their recent origin, nations had become very real places to citizens through politics and the propaganda machinery of national governments. On the other hand, regionalism was only promoted by a few writers and artists, and was unlikely to impress broad layers of the population unless regions assumed political importance. However, since then, sub-national as well as macro-regions have attracted the increased attention of political elites. Little empirical research has investigated how macro-regions are perceived outside academic and political circles.

Such studies alone can determine what regional narratives mean for the wider public, and the extent to which they comply with region-building images produced by elites. This article examines the mental maps of high school seniors in 10 cities in the Baltic Sea and Mediterranean regions.

Seas and waterways are generally perceived as constituting the borders between countries or regions, although there have been historical periods when central seas have served as means of unification. In recent decades, several concurrent political and technological developments have had the potential to alter the image of the Mediterranean and the Baltic Sea as barriers on mental maps. Whether this has in fact taken place remains an open question. There is no doubt that such a development would be considered desirable by political leaders and scholars, who have tried by various means to bring about the consolidation of regions around these seas. History has traditionally been written within a national framework, but the appearance *La Méditerranée et le monde méditerranéen à l'époque de Philippe II* by Fernand Braudel in 1949 marked the beginning of regional history writing, and similar attempts were made with regard to the Baltic Sea region by scholars such as Klinge (1995) and Gerner *et al.* (2002) after the Cold War. On the political level, regional organizations and parliaments such as the Euro-Mediterranean Partnership, the Parliamentary Assembly of the Mediterranean, and the Baltic Sea Parliamentary Conference have been increasing their activity over the past few decades, and have dealt with questions of trade and environmental protection. The attempts to revive a medieval trading organization through the New Hansa were an effort to combine historical and economic region building. All of these are attempts to foster a sense of regional connection united, rather than divided, by the large seas at their core.

By surveying students in their last year of secondary school in the Baltic Sea and the Mediterranean regions, we hope to establish whether the Mediterranean and the Baltic Sea serve to connect or divide regional perceptions on students' mental maps. We used a survey to determine student knowledge and attitudes regarding different countries in these regions.[1] By comparing the survey results from 2014 to 2015 with similar studies conducted in the 1960s, the present study also attempts to ascertain whether attitudes toward countries in those regions have changed in the intervening period.

Mental maps

The concept of a "mental map" has been introduced to describe how people orient themselves within their environment and how they perceive the world, and is approximately synonymous with the idea of cognitive maps. Although both terms are used in geography, behavioral science and psychology, the term "mental maps" is more common among geographers (Hannes *et al.* 2012). The term mental map has recently also been used by historians in attempting to describe the worldviews of political leaders (Casey and Wright 2008, 2011), the French colonial mind (Thomas 2011), or images of the region around the Baltic Sea (Götz *et al.* 2006). Little communication has so far taken place between those engaged in the historical and geographical research on mental maps. The elaborate methods and definitions developed by geographers have not reached historians, and the interest of historians in how political developments contribute to changes in mental maps has so far not reached geographical researchers. This article will attempt to combine the strengths of both approaches to mental mapping by studying the historical question of change over time using methods developed in the field of geography.

The term mental map might refer to one of two things: the images of the surrounding world that people hold in their minds, and an actual map which a researcher has constructed in order to represent the worldview of his research objects. The term is used here to describe maps the author has constructed on the basis of survey data. These maps do not represent the mental map of a single individual, but an average of the maps of a group of people.

The pioneer of the survey method in mental map research was the geographer Peter Gould. From the mid-1960s on, he investigated student knowledge and perception of areas such as US states (1966), municipalities in Sweden (1975a), and countries in Europe (1966) by constructing information and perception maps. While Gould recognized that his definition and measurement of information was grossly simplified, he claimed that such simplification was inevitable at the beginning, but could lead to the exploration of more complicated questions (1975b, p. 88).

Gould believed that the mental maps of those living in a certain place resulted from information flows. The closer a place was, and the larger its population, the more information people received about it and the more prominent it became on their mental maps. However, barriers to information flow (political, such as national borders, or natural, such as mountain ridges or waterways) could alter this picture (Gould and White 1974, pp. 141–143). When launched, the model seemed very simplified. For example, it ignored Galtung's (1971) suggestion that the direction and strength of global information flow are also affected by post-colonial center–periphery relations. Today, in the information age, the view of individuals as passive receivers of information flow seems even more oversimplified, as it is likely that the importance of active information-seeking has increased. Since Gould conducted his research, information-seeking behavior has become a well-established field of research within library and information science (Case 2002).

Gould also found that people living in different locales generally agreed on which places they considered attractive to live in, with the exception of the "local dome of preference", i.e. the particular attractiveness of nearby locations. However, Gould found that not all of their neighbors were positively perceived. The great quantity of information respondents had about nearby locations enabled them to better discriminate between nearby places than a distant observer was able to do. For example, people in the US living in the South were found to draw strong distinctions between positively and negatively perceived southern states, all of which seemed equal to observers from the North (Gould and White 1974, p. 100).

Mental mapping research experienced a peak from the 1960s to the 1980s, after which even Gould had exhausted his interest in the field. However, it is relevant to revisit his methods today because the geographers who conducted mental mapping research in that era primarily sought to discover general principles and regularities in mental maps. From the point of view of a historian, their empirical results are the mental maps of a particular period in time, shaped by its special political and social circumstances. In that respect, old mental mapping research is an important source to help us understand the worldviews of the 1960s, for example. When we do similar studies today, we can discover how present mental maps differ from older ones, and inquire into how and why mental maps change. Issues of how permanent mental maps are, and how they are transformed by historical developments, remain questions barely addressed by the first generation of mental mapping researchers.

Waterways on mental maps

Several writers in the field of mental mapping have discussed the functions of seas and other waterways on mental maps. Lynch (1970) observed that the strongest edges – linear elements usually forming boundaries between areas – on peoples' images of cities were visually prominent and difficult to cross. Many of the examples he cited were rivers or waterfronts. In a survey of 10,000 Swedish schoolchildren, Gould found that lakes were effective barriers to information flow as, for example, in Skövde, located between the two large Swedish lakes of Vänern and Vättern (1975a, p. 84). The area on the opposite shore of Lake Vättern was as unknown to the children as the interior of northern Sweden. Hägerstrand (1953) also found that lakes were contact barriers in rural Swedish districts. He attributed this to the dominant role that roads have played since the advent of automobiles, although he believed that lakes still had a connecting function in the early 1900s.

Also, a remote sea might serve as an important divide between geographical regions on mental maps. When Mozambican Islanders were asked to draw world maps, the Mediterranean was the sea most represented. Madaleno interpreted this as reflecting the knowledge that the Mediterranean separates Europe from Africa (2010, p. 119). The Eurobroadmap project, based on a worldwide survey of 9000 students in 18 countries, also found that the Mediterranean (especially its western parts) formed a sharp divide between perceived regions on mental maps (Didelon *et al.* 2011). Geographical research on the effects of waterways on mental maps has gained good insights into the present situation by utilizing survey instruments or quantitative methods, but they have not investigated changes in mental maps over time. However, like Hägerstrand, individuals have speculated as to what the situation might have looked like in the past. Although the mental map concept has not achieved the same level of theoretical understanding and methodological development among historians as it has among geographers, historians have attempted to investigate changes of mental maps to a greater degree.

According to Jackson (2009), medieval Scandinavians divided the world into four quarters: Sweden and all the lands to the east of the Baltic Sea on the route to Constantinople belonged to the eastern quarter, called Austrvega. In the thirteenth century, however, the designated region was narrowed to the lands on the South and East coasts of the Baltic Sea. Thus, the Baltic Sea was a central waterway for the eastern quarter of the world during the maritime early middle ages, but it became a dividing line between quarters with the advent of earth-bound feudal society.

A similar shift was seen in the early 1800s, when the map of Northern Europe was redrawn after Sweden lost Finland to Russia and gained Norway from Denmark (Berg 2014). In the process, the Baltic Sea, Skagerak, and Kattegatt were transformed from central hubs of the Nordic monarchies to defensive borders. In order to gain legitimacy for the change, rulers and scholars stressed the "naturalness" of the new borders, reinterpreting the mountain ridge that separated Sweden and Norway as a unifying "flowering meadow".

Earlier research confirms that the role of waterways on mental maps has shifted during political, technological, and social revolutions. This raises the question of how the developments of recent decades – such as the fall of the Iron Curtain, region-building efforts, and advances in information technology – have influenced the mental maps of the Baltic Sea and Mediterranean regions.

Research question

When the Cold War ended in the early 1990s, the Baltic Sea ceased to be the nearly impenetrable Iron Curtain it had been since the late 1940s. In the 1990s, scholars and politicians made attempts to strengthen the regional identity of the Baltic Sea region by talking it into existence (Götz 2016). More than two decades later one might have expected that region-building efforts in the area would have ended or at least weakened the Baltic Sea's role as a sharp divide between positively and negatively perceived places.

Attempts at region building have also been made in the Mediterranean, but with limitations regarding free movement. While the Mediterranean continues to function as an avenue for immigrants and refugees fleeing conflict zones, European nations have tried to reinforce the sea's properties as a barrier between continents by, for example, patrolling the sea and targeting migrant smugglers. Since the turn of the millenium, nationalistic and ethnocentric populist movements have gained support globally, including in the countries surveyed. By depicting foreigners as others against whom they want to draw distinct barriers such political movements has probably affected mental maps of foreign countries and people on the other side of the sea. In recent years, the aftermaths of the Arab Spring and tensions in Eastern Europe have also disturbed region-building in the Mediterranean and Baltic Sea regions.

The past few decades have also seen an increase in ferry and air traffic, enabled in the Baltic by the fall of the Iron Curtain, and in the Mediterranean spurred on by tourism and the rising number of North African expatriates in Europe, who frequently travel between their new and old homelands. Global sea transportation has quadrupled from 1992 to 2012 (Tournadre 2014). An even more revolutionary development has taken place in the field of electronic communications. According to Sassen, the Internet heralded the beginning of a post-national world in which physical borders and distances are of less importance than ever before (2002, p. 277). If correct, this would alter the degree to which seas are perceived as barriers. Thus, both theoretical arguments and parallels to the findings of earlier research make it likely that these developments have affected the mental maps of the two regions as held by their inhabitants. However, exactly what these mental maps look like after these transformations can only be determined through an empirical investigation.

The following pages will first examine to what extent the Baltic Sea and the Mediterranean function as either barriers or facilitators of information flow, based on the knowledge the people living around those seas have about areas on the opposite shore. Second, what people think about countries situated around the mutual sea that adjoins them will be investigated, including whether neighboring countries are incorporated in a regional dome of preference, regarded with indifference, or negatively perceived. We will also consider whether a change has taken place in such preferences since Gould surveyed the image of Europe among Swedish and Italian students in the 1960s.

Method

The results of the present study are represented by geographical information and geographical perception maps based upon data from surveys. Thus, *information* and *perception* are here used in a simplified way, to refer to the quantitative data that students produced in answers to the questionnaire described below.

The questionnaire and maps

The *geographical information* map is based on the first question: the respondents were given a blank sheet of paper and asked to, in five minutes, write down as many place names as possible, freely chosen without geographic limitations as to scale, type of place, or region. We then coded these place names by country, and constructed choropleth maps based upon that country's percentage among all the place names mentioned.

The *geographical perception* maps are based on the second question: students were asked to rank each country on a blank map that included the areas in Figure 4 on a scale from one to five, with five indicating areas where they would most want to live. The map encompassed Europe and North Africa, thus including the Baltic Sea and Mediterranean regions. Based on the average of the points awarded each country, maps were constructed illustrating the perception of the students. Countries which were ranked by fewer than five students in a class were left blank on the map.

In 1966, Gould conducted a study of the perception of Europe among university students in European cities, including Uppsala and Rome. Using a map of Europe, he asked the students to order the countries beginning with the one in which they would most like to live. The existence of Gould's study enables a comparison of how the perception of Europe in Italy and Sweden has changed from 1966 until the present. However, since the political map of Europe has altered since 1966, it is not possible to exactly duplicate Gould´s method. Therefore, a method suggested by Thill and Sui was used (1993). They have argued that ranking countries on a scale from 1 to 5 captures the middle range of moderate preferences more exactly than ranking 40 countries in order from 1 to 40, as in Gould's surveys. However, Thill and Sue concluded that the results of both methods were similar, which makes a comparison of this study's results and those of Gould meaningful.

The selection of schools and the conduct of the surveys

The schools surveyed were chosen to enable an investigation of the mental maps of the Baltic Sea and Mediterranean regions held by students living in different locations in relation to these seas. Since the purpose of the study was to investigate to what extent the Mediterranean and Baltic Seas function as barriers on mental maps, we surveyed students in locations on both sides of the former Iron Curtain in the Baltic Sea and on both sides of the continental divide between Europe and Africa in the Mediterranean. We selected schools to ensure a variety of locations in relation to the sea in question – inland, coastal, and island locations. Since differences between central and peripheral locations within each nation might affect the results but was something we did not intend to study, capitals were avoided. A total of 10 locations were selected: one on an island in the middle of each of the seas as well as two to the north and south for the Mediterranean or east and west for the Baltic Sea region, one on each side of the sea, one on the coast, and one inland: Fez and Tangier in Morocco, Malta, Venice and Bologna in Italy, Uppsala and Gävle in Sweden, the Åland Islands (belonging to Finland), and Noarootsi and Valga in Estonia. The inclusion of Swedish and Italian locations enables a comparison with Gould's 1966 survey.

Differences in school systems among the countries surveyed complicated the study. Educational diversity within each country has also increased over the last few decades, as schools have striven to develop specialized profiles with, for example, language or sports classes. However, the strong similarities between this study's maps of Swedish students, for example, and those found in the Eurobroadmap project (Grassland 2012), which surveyed a much larger number of students, suggest that the present study surveyed classes that were typical. However, the primary aim of this study was not to determine national representativeness, but to investigate the effect of *different locations* on the mental maps of the Baltic Sea and Mediterranean regions. Despite diversity, the basic similarity of school institutions worldwide makes them ideal for comparisons between countries as different as, for example, Estonia and Morocco.

With regard to age differences, students in Malta have already begun college at an age when other students in the survey are in the last year of secondary school. Therefore, Maltese first-year college students were surveyed. On Åland, students in the junior year of secondary school were surveyed, since those in their senior year were preoccupied with final exams. In Noarootsi, last-year students were combined with first-year students in order to get a sufficiently large sample (Table 1). Our sample size, one class at each location, is similar to the one used in many previous mental mapping research (such as Gould 1966). That the findings of these smaller studies are congruent with the ones based on much larger samples, such as Eurobroadmap or Gould (1975a), does indicate that a class size sample is sufficient in order to provide relevant results.

Results

Geographical information

Students from the western European mainland (Italy and Sweden) displayed little knowledge of countries on the other shore of their bordering seas. The most extreme example was in Uppsala, where no one cited a single Russian place name. Sweden's neighbor, Finland, was better known than Russia and the Baltic countries, but Uppsala students still knew more about Spain, Turkey, and France than they did about Finland (Figure 1).[2]

Table 1. Classes surveyed.

Place	Date	Size	Grade in upper secondary school	Surveyed during lesson in	Survey language	Profile/ specialization of school or class
Valga, Estonia	1 Apr. 2015	30	3 of 3	Estonian	Estonian	
Noarootsi, Estonia	2 Apr. 2015	19	1 and 3 of 3	Swedish	Estonian	Boarding school, Scandinavian profile
Åland. Finland	21 Mar. 2014	18	2 of 3	Geography	Swedish	
Gävle, Sweden	2 June and 25 Aug. 2014	31	3 of 3	History	Swedish	Cultural (theater, dance, circus)
Uppsala, Sweden	23 Sept. 2014	25	3 of 3	History	Swedish	Social science program
Bologna, Italy	11 Nov. 2014	23	5 of 5	History	Italian	Polytechnical school (Istituto tecnico)
Venice, Italy	12 Nov. 2014	16	5 of 5	English	Italian	Language school
Valetta, Malta	16–17 Nov. 2015	31	Junior college	Geography (on a break)	English	
Tangier, Morocco	25 Nov. 2015	18	3 of 3	English	Arabic	Science class
Fez, Morocco	23 Mar. 2015	41	3 of 3	English	Arabic	

Students in Italian schools knew little about the southern and eastern parts of the Mediterranean region. France and Spain (3–6% of the total number of place names) were the only Mediterranean countries about which they had substantial information (Table 2).

Figure 1. Uppsala information map displaying each country's percentage of the total place names mentioned by students.

Table 2. Top 10 countries according to their percentage share of the total number of place names mentioned by students in each location.

Uppsala	%	Gävle	%	Åland	%	Noarootsi	%	Valga	%
Sweden	81.0	Sweden	69.0	Sweden	21.0	Estonia	50.0	Estonia	32.0
USA	4.2	USA	6.4	Finland	15.0	USA	4.0	USA	5.4
Spain	3.0	Germany	2.4	Åland	14.0	Sweden	3.1	Latvia	4.5
Turkey	1.7	UK	2.1	USA	7.2	Russia	2.7	Germany	3.4
France	1.1	Italy	2.0	UK	5.4	Finland	2.4	Russia	3.3
Finland	1.0	France	1.9	Germany	3.4	Italy	2.1	Italy	3.2
Egypt	0.8	Norway	1.8	France	3.2	Latvia	2.2	Finland	2.9
Greece	0.8	Denmark	1.2	Italy	2.8	UK	1.8	Lithuania	2.7
Italy	0.8	Spain	1.2	Norway	2.5	Germany	1.8	UK	2.1
Norway	0.7	Greece	1.1	Russia	2.3	Norway	1.6	Sweden	2.1
Bologna	%	*Venice*	%	*Malta*	%	*Tangier*	%	*Fez*	%
Italy	57.0	Italy	20.0	Malta	21.0	Morocco	81.0	Morocco	20.0
USA	5.8	USA	9.8	Italy	10.0	France	1.8	France	4.2
Spain	4.1	Spain	5.7	USA	6.7	Germany	1.3	USA	3.8
France	3.1	France	4.8	UK	6.1	USA	1.3	Spain	3.1
UK	2.6	UK	4.0	France	4.3	Spain	1.1	Germany	2.6
Germany	1.6	Australia	2.7	Germany	3.5	UK	1.1	Italy	2.6
China	1.2	Canada	2.6	Spain	2.4	Tunisia	1.1	Saudi Arab.	2.6
Portugal	1.1	China	2.5	Australia	2.0	Turkey	1.1	UK	2.5
Russia	1.1	Japan	2.4	Libya	1.7	Algeria	0.8	Sweden	2.4
Ukraine	0.9	Russia	2.1	Japan	1.6	Russia	0.8	Turkey	2.3

Students in the two Estonian schools surveyed had a more detailed information map of Sweden than the Swedish students had of the Baltic countries. Likewise, the Moroccan students had more information of the northern shore of the Mediterranean than the Italians had of the southern shore. Moroccan students tended to have much more information about European countries in the western Mediterranean than about their North African neighbors. Their information maps show the North–South divide to be less important than the East–West divide.

In the island locations surveyed, Åland and Malta, knowledge of both shores of the sea was relatively good. Students on Åland had more information about Sweden than the Estonians and more information about all other countries around the Baltic Sea than the Swedes. Maltese students knew more about nearby Italy than they did about any other foreign country. In addition, Maltese students displayed significantly more knowledge of nearby Libya than any other group of students in the survey, and they knew a great deal about other Mediterranean countries as well. The Maltese also displayed considerable information about Cyprus, which, outside of Malta, was only mentioned by one student from Åland. However, the Ålanders' and Malteses' relatively good knowledge about the countries surrounding the two seas seems to be a reflection of their broader knowledge of maritime locations in general, and was not limited to the Baltic Sea and Mediterranean regions. This is illustrated by the example of Britain. Unlike the Italian students, the Maltese had more information about Malta's former colonial ruler, England, than about France and Spain. Students on Malta and Åland knew significantly more about England than students in any other location surveyed. As the central metropolis of a former maritime empire, England has for centuries been in contact with these island communities, which seems to have left a lasting impression on mental maps.

In both Sweden and Estonia, students from coastal schools knew more about the opposite shore than students from inland schools. The pattern that the other shore of the sea is better known in coastal locations holds true for Italy, where Venetian students had more detailed information maps of North Africa than Bolognese students. However, an exception to this rule was Tangier, where students had much less information about Spain than their counterparts from Fez, which is all the more surprising since under favorable conditions Spain can even be seen from Tangier. Only 1.6% of the place names mentioned by students from Tangier were Spanish, compared to 3.1% from Fez, 5.7% from Venice, and 3% from Uppsala. Some of the Spanish place names mentioned by Moroccan students are not situated north of the Straits of Gibraltar; Ceuta and Melilla, enclaves in Morocco, constitute one-third of the Spanish place names cited in Tangier and one-fifth in Fez. The area around the Gibraltar Straits, which encompasses a continental divide and multiple enclaves, might be described as an unusually complex borderscape. According to Rajaram and Grundy-Warr (2007), by excluding the chaotic outside, borders instill a sense of community, and in the process may produce hidden geographies. It is possible that the location of Tangier close to an ambiguous political border creates the need for a clear and distinct mental border, separating a well-known "us" from a "them", virtually forming a hidden geography. However, it has to be remembered that the Tangier sample is the least robust in this study, since class size was relatively small and the number of place names mentioned per student was low, contributing to a low total of placenames (Tables 1 and 2).

In summary, although the other shore of the sea was better known to those east of the Baltic Sea and south of the Mediterranean, and also on the islands, than in the western European mainland west and north of these seas, those bodies of water still seem to serve as barriers to information, albeit to a different degree depending upon where one resides. Thus, he information maps are not characterized by good knowledge about countries around the Baltic and Mediterranean seas. Instead, they are dominated by other traits.

The geographical information maps appear to be predominantly national. In Uppsala and Tangier, more than 80% of the place names mentioned by students were located in Sweden or Morocco, respectively. The other extreme was Venice, where only 20% of the place names listed by students were Italian. All classes had more than double the amount of information about their own nation than about any other country, with the exception of Åland, where Swedish place names constituted 21%, mainland Finland

15%, and Åland 13% of the total. This pattern might well be related to the island's historical and geographical location between Sweden and Finland; it is a monolingual Swedish-speaking autonomous province of Finland, which until 1809 was a part of Sweden.

While maps from most locations focused upon Europe and the US and left large parts of the globe (especially Africa) blank, almost all continents were well represented on the mental map from Fez (Figure 2).

According to Galtung's (1971) theory, the main direction of global information flows is from the center of the West to the periphery of the developing world, while much less information flows in the opposite direction. This would explain why the periphery, such as Fez, has more elaborate mental maps of the center than the other way around. However, another of Galtung's hypotheses, that there is little or no flow of information between peripheral countries, is contradicted by our finding that the most detailed mental maps of the periphery were, in fact, to be found in the periphery. This pattern is confirmed by data from the Eurobroadmap project.[3] For example, it showed that Cameroonian students had much more elaborate mental maps of the peripheral parts of the world than Swedish students.

Geographical perception

In Gould's investigation of residential desirability in Europe from 1966,[4] Swedes considered Greece and Portugal (with their then-authoritarian regimes) as unattractive as the socialist countries, while Italians perceived Greece more favorably. However, Italians ranked Finland, which Swedes almost placed on a par with Sweden, poorly. Gould interpreted the Italians'

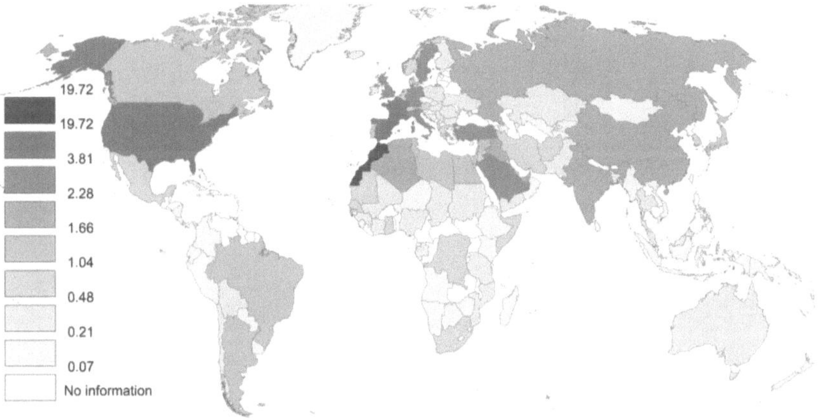

Figure 2. Fez information map displaying each country's percentage share of the total number of place names mentioned by students (1443).

positive view of France, Spain, and Portugal as a preference for "Latin" countries, but since the positive view also encompassed Greece it might as well be described as a Mediterranean preference (Figure 3).

Our 2014–2015 survey makes it possible to investigate whether the Mediterranean preference of Italians have been strengthened in the intervening years, and whether the positive Swedish view of Finland, after the fall of the Iron Curtain and decades of attempted Baltic Sea region building, has developed into a general preference for countries in the region. From a region-building perspective, the results are discouraging. In fact, the trend has been in the opposite direction. Notably, in 2014 Finland was more positively perceived in the Italian cities of Bologna and Venice than in the Swedish cities of Gävle and Uppsala (Figure 4). This also represents a change from 1966, when Gould found that Swedish university students had a more positive view of nearby Finland than students from other countries in Europe, including Italy (1966, pp. 27–51). A similar change has taken place regarding Sweden's other next-door neighbor in the Baltic Sea region, Denmark. In 2014, other than Italy itself, only two countries in the Mediterranean region (Spain and Greece) were perceived more favorably in Italy than in Sweden, and that only by a slight margin. Thus, Italians and Swedes seem to have experienced negative region building since the 1960s: today they are not particularly fond of their neighbors and would prefer to live in each other's regions rather than living in their own. This is clearly illustrated by Figure 4.

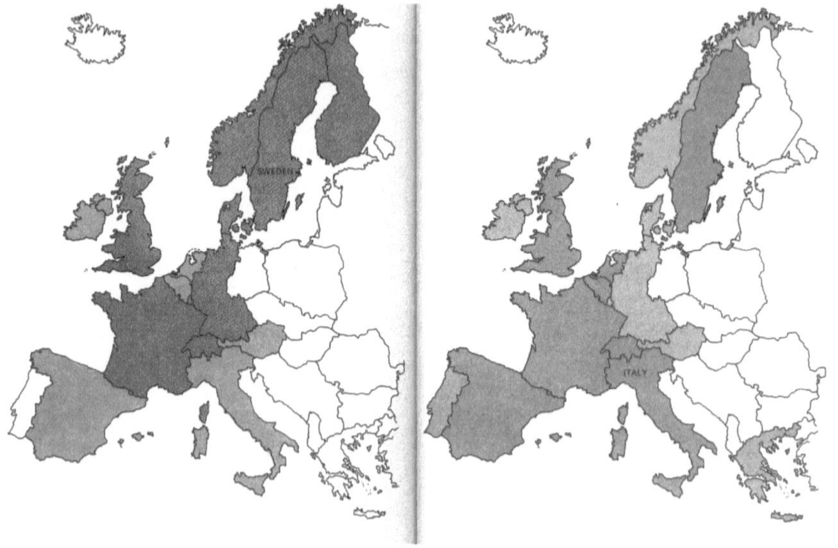

Figure 3. Residential desirability of European countries in 1966. Darker colors mean more positive perception. Perspectives of university students from Uppsala (left) and Rome (right). (Based on Gould and White (1974, pp. 182–183).)

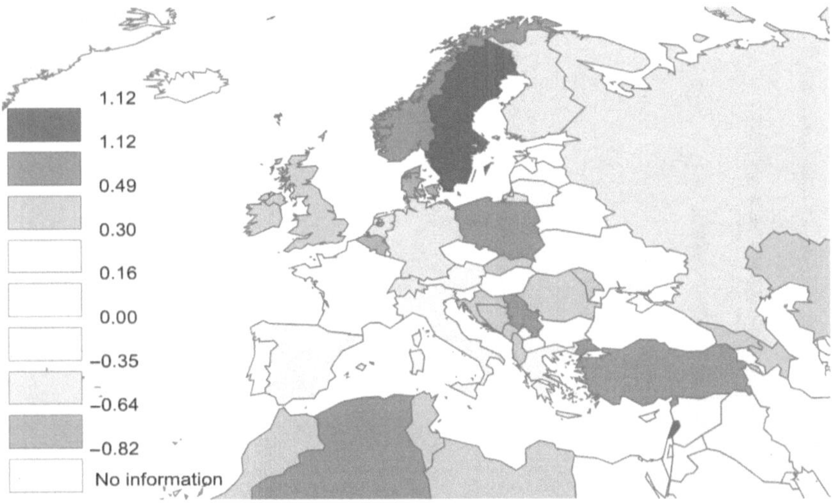

Figure 4. Aggregate map showing differences in perception between Swedish and Italian secondary school students in 2015. Map is constructed by subtracting averages in perception awarded by Italian students (Bologna and Venice) to each country on the map from averages as rated by Swedish students (Uppsala and Gävle). Red colors are more positively perceived in Sweden, blue in Italy.

This phenomenon is not limited to Swedes and Italians, but can be found in the other locations surveyed as well. The perception maps from those locations surveyed in 2014–2015 share the common feature that the countries on the eastern and southern fringes of the map are generally more negatively perceived, although the south is most negatively perceived by Mediterranean students and the east by students from the Baltic Sea. For example, in all the locations surveyed in the Mediterranean, students indicated they would prefer living in Russia to living in Algeria, while students from the Baltic Sea region preferred Algeria to Russia (Figure 5). This is the opposite of what one might expect if region building had been successful in building a regional consciousness and identity among residents around the Baltic and Mediterranean seas. It indicates that the rivalry among neighbors is stronger than local preference in these regions.

Within the Mediterranean region, the negative perception noted above cannot simply be explained by a sharp North–South divide between African and European countries. In 1966, most Italian students had a relatively positive view of nearby Albania, which they valued almost as highly as Finland. On the other hand, Albania was ranked at the bottom of the charts in the rest of Europe, including Sweden. In 2014, however, Albania was more negatively perceived in Italy than in Sweden. In both Bologna and Venice, all North African countries were perceived more favorably than Italy's neighbor, Montenegro. In Venice, Albania was perceived slightly more favorably than Algeria and Tunisia, but was ranked lower than all

A colour version of this figure is available in the original journal article.

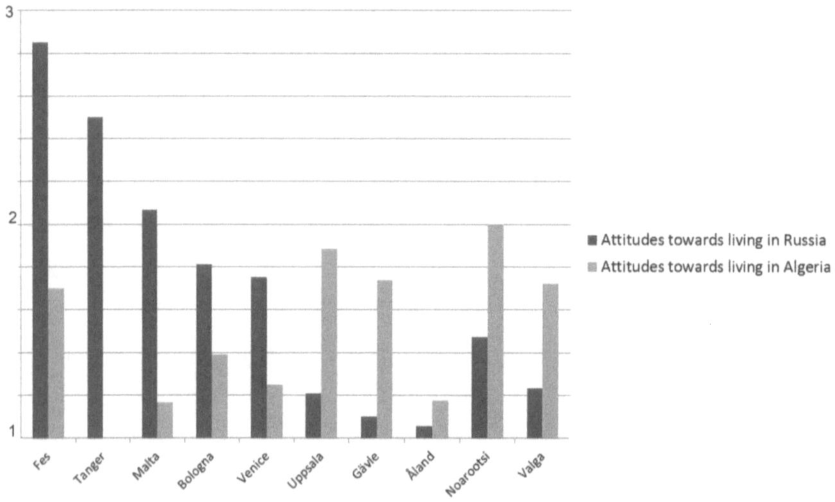

Figure 5. Average perception of Russia and Algeria by students from locations surveyed.

countries in North Africa by Bologna students. In Venice, Egypt and Morocco were considered more popular places to live than Croatia. These examples indicate that the Mediterranean Sea does not function as a simple fault line between North and South in the mental maps of Italian secondary school students, but that the picture is more complex. Moreover, the fall of the Iron Curtain has not automatically lead to a more positive perception of those areas that were once sequestered behind it.

That the Mediterranean is perceived positively in the Baltic Sea region is not surprising. Already Saarinen (1973) noted a "nostalgia for the balmy Mediterranean" on the mental maps of Finnish students, and believed it might be an example of how residents of northern climates looked in general upon areas with more benevolent climate. Another part of this study (to be published in a forthcoming article), where we asked students about their overall perceptions about the Baltic and Mediterranean regions, confirmed that climate was perceived as one of the great advantages of the Mediterranean. Climate might also explain the fact that Sardinia and Corsica appear on the list of the top ten most preferred "countries" of residence in Gävle and Valga (Table 3). That many Mediterranean islands are perceived as countries by the student's is probably related to their insular position as separate geographical entities delimited by their coastlines. Since the ranking was conducted on a blank map, it is possible that the students did not know exactly what islands they were ranking, but awarded them high points anyway based on their attractive location in the Mediterranean.

Nevertheless, this northern gaze on the Mediterranean climate seems to be a constant background factor and cannot explain the changes in mental maps

Table 3. Top 10 preferred countries as rated by students.

Uppsala		Gävle		Åland		Noarootsi		Valga	
UK	4.3	UK	4.6	UK	4.7	Estonia	4.3	Italy	4.3
Italy	4.1	Sweden	4.3	Sweden	4.4	Italy	4.3	UK	4.2
France	4.1	France	3.9	Norway	4.3	UK	4.2	Estonia	4.2
Sweden	4.0	Sardinia	3.8	Germany	3.9	Sweden	4.2	Germany	4.0
Spain	3.9	Italy	3.7	Denmark	3.9	France	4.1	France	4.0
Portugal	3.3	Ireland	3.6	Finland	3.6	Ireland	3.8	Spain	3.8
Norway	3.2	Spain	3.6	France	3.6	Germany	3.8	Portugal	3.8
Switzerland	2.8	Norway	3.4	Italy	3.6	Denmark	3.7	Netherlands	3.5
Germany	2.7	Germany	3.2	Ireland	3.5	Spain	3.7	Corsica	3.5
Ireland	2.6	Portugal	2.9	Spain	3.5	Netherlands	3.6	Ireland	3.5
Bologna		*Venice*		*Malta*		*Tangier*		*Fez*	
Spain	4.3	UK	4.5	Malta	4.5	Sweden	4.4	Netherlands	4.6
Italy	4.2	Spain	3.9	Germany	4.0	Morocco	3.8	Norway	4.6
France	4.2	Germany	3.7	Mallorca	4.0	Turkey	3.7	Sweden	4.3
UK	3.8	Italy	3.7	UK	4.0	Germany	3.4	Ireland	4.2
Germany	3.5	Denmark	3.5	Ireland	3.9	UK	3.3	UK	4.0
Portugal	3.2	France	3.4	Norway	3.8	France	3.2	Turkey	3.9
Sweden	3.0	Netherlands	3.1	Finland	3.8	Spain	2.2	Morocco	3.7
Switzerland	2.9	Sweden	3.1	Switzerland	3.7	Algeria	1.0	Germany	3.5
Belgium	2.8	Switzerland	3.0	Italy	3.6			France	3.4
Greece	2.7	Ireland	2.9	France	3.6			Saudi Arab.	3.3

Note: Students ranked countries on the map according to residential desirability on a scale from 1 (lowest) to 5 (highest). In Tangier, only eight countries were ranked by more than five students and included in the table. Some sub-national regions, mostly islands in the Mediterranean, were also ranked as countries by some students.

that have taken place since the 1960s. It seems, rather, that rising political tensions and nationalist tendencies in the regions may have brought about an atmosphere of distrust among neighbors.

Conclusions

Our study shows that, despite the end of the Cold War, attempts at region building in the Baltic Sea and the Mediterranean, and increased communication, traffic, and massive migration, the Baltic Sea and the Mediterranean continue to function as borders on mental maps. Most students surveyed had little information about countries on the opposite shore of their neighboring sea, although such knowledge is comparatively better among students from coastal – and particularly island – locations.

The students surveyed from 2014 to 2015 did not display a local preference for nearby countries within their region. While Swedish and Italian students did so in the 1960s, today's students seem to prefer the other country's neighbors to their own. There is a strong tendency to discriminate between countries in one's own region, as reflected in the negative perception of Russia in the Baltic Sea region, and North Africa and the Balkans in the Mediterranean.

Since the 1960s, the tendency to perceive neighbors negatively has come to dominate the earlier tendency for local preference, at least in Sweden and Italy, where direct comparisons can be made. It is plausible that the atmosphere of crisis, threat, and uncertainty at the time of our surveys (2014–2015) contributed to the prevalent negative attitudes toward neighboring countries. Thus, our quantitative study of mental maps, using concepts and methods from the field of geography, seems to confirm the conclusion earlier drawn by historians, based on textual analyses, that mental maps are sensitive to rapid political changes.

We found that mental maps were in most cases overwhelmingly national. Hopes that the Internet would usher in a new post-national world, just as the printing press paved the way for nationalism, were not confirmed by this study. In the words of Catherine Frost, the anonymous, flexible, low-commitment community on the Internet has thus far not been able to "generate the kind of boundaries and shared meaning that are required for a new system of social and political solidarity" (2006, p. 47).

If the information society had altered mental maps in a direction away from the local and regional, and toward a more global worldview (in which physical distance and location has little importance), we would expect to have found elaborate geographical information maps depicting all corners of the world, particularly among students from countries like Sweden, where there has long been extensive penetration of the Internet and other electronic media. However, this does not correspond with our actual findings:

Swedish geographical mental maps were extremely national, while the most dispersed geographical information map was found in Fez, Morocco. Nor can the comparatively detailed mental maps of Africa found in Fez be explained by Galtung's theory of information hierarchies, according to which little information should pass from one Third World country to another.

Our findings are difficult to explain without recognizing that mental maps are not simply based upon the reception of information from the surrounding world, but upon an active search and interpretation of information. Reflexes of information selection, such as the strength of one's national or Eurocentric preconceptions, seem more important in determining the place names one learns – and retrieves from memory in a survey situation – than potential access to information.

Such preconceptions have probably formed over generations through interplay between spatial and historical factors, as well as deliberate attempts at identity construction. National and Eurocentric identity conceptions have managed to profoundly influence the mental maps of the students in most locations surveyed, while the Baltic Sea and Mediterranean region-building projects have not. In these two regions, Yi-Fu Tuan's 1975 prediction that, unless it assumed political importance, regionalism was unlikely to impress broad layers of the population still holds true.

Since mental maps are not simply based upon the reception of information from the surrounding world, but upon an active search and interpretation of information, seas like the Baltic and the Mediterranean may continue to function as barriers on mental maps, despite the fact that, thanks to modern information technology, they are no longer barriers to information flow in the direct sense of the word. Since students East of the Baltic and South of Mediterranean Seas had more information about the opposite shore than the other way around, Galtung's theory of global information flows might offer an explanation for this continued barrier function. If information flow from periphery to center is weak, as Galtung suggests, this will affect the mental maps in central countries, such as Sweden and Italy, by impairing knowledge of countries at the opposite shore. It is probably a hindrance to the formation of elaborate mental maps of the Baltic Sea and Mediterranean regions that these seas are surrounded by countries which have very different positions in the center–periphery system of information flow.

By revisiting old mental mapping research and comparing it with new survey results, this study has shown that mental maps do undergo significant changes over time, and that these changes are likely associated with political transformations. However, ours is only a first foray into a field that is largely unexplored, and the interpretation of our findings is hampered by the long time period (1966–2014) that has elapsed between the surveys. The major political events that have taken place in the interim make it difficult to link the changes in mental maps to any specific occurrence. For example, it is not impossible that the fall of the Iron Curtain did improve perception of countries east of the Baltic Sea, but this effect may be undetectable in 2014, since it was temporary and has been reversed by later developments, such as the war in Ukraine and rising tensions in the Baltic region. To more systematically investigate how stable mental maps are, and how they are affected by political events, would require further mental mapping research, carried out at more frequent intervals.

Notes

1. The survey also investigated knowledge and attitudes about the history of these countries. These aspects are analysed in Holmén (2017).
2. The limits between the eight colors on the information maps have been set according to the Jenks distribution, which allows optimal differentiation on choropleth maps. This distribution sets divides independently for each map; thus, colors do not represent identical values on all maps.
3. Mental maps from the Eurobroadmap project surveys can be viewed using the online tool "subjective mapper" at http://www.ums-riate.fr/mapper/.
4. The study is described in Gould (1966) and in Gould and (White 1974). However, principal component analysis was used in 1966, and the 1966 illustrations of Italian and Swedish students are based on the first dimension found in the analysis. The illustrations reprinted here are from 1974 and encompass the entire group of students.

Acknowledgements

I would like to thank my colleagues in the project Spaces of Expectation; participants in the workshop on "Mental Maps: Historical and Social Science Perspectives" held in Stockholm, 12–13 November 2015; and members of the seminar on History of Education at Uppsala University for all of their valuable input.

Disclosure statement

No potential conflict of interest was reported by the author.

Funding

This work was supported by the Foundation For Baltic And East European Studies.

ORCID

Janne Holmén http://orcid.org/0000-0003-2449-4888

References

Berg, R., 2014. Denmark, Norway and Sweden in 1814: a geopolitical and contemporary perspective. *Scandinavian Journal of History*, 39 (3), 265–286.

Braudel, F., 1949. *La Méditerranée et le monde méditerranéen à l'époque de Philippe II*. Paris: Armand Colin.

Case, D.O., 2002. *Looking for information: a survey of research on information seeking, needs, and behavior*. San Diego: Academic Press.

Casey, S. and Wright, J., eds., 2008. *Mental maps in the era of two world wars*. Basingstoke: Palgrave Macmillan.

Casey, S. and Wright, J., eds., 2011. *Mental maps in the early cold war era, 1945–1968*. Basingstoke: Palgrave Macmillan.

Didelon, C., *et al.*, 2011. *Mental maps of students*. Vol. 5 [online]. Available from: https://halshs.archives-ouvertes.fr/halshs-00654530 [Accessed 23 May 2015].

Eurobroadmap subjective mapper [online]. Available from: http://www.ums-riate.fr/mapper/ [Accessed 27 Jan 2016].

Frost, C., 2006. Internet galaxy meets postnational constellation: prospects for political solidarity after the Internet. *The Information Society*, 22 (1), 45–49.

Galtung, J., 1971. A structural theory of imperialism. *Journal of Peace Research*, 8 (2), 81–117.

Gerner, K., Karlsson, K., and Hammarlund, A., 2002. *Nordens Medelhav. Östersjöområdet som historia, myt och projekt* [The Nordic Mediterranean: the Baltic Sea region as history, myth, and project]. Stockholm: Natur och kultur.

Götz, N., 2016. Spatial politics and fuzzy regionalism. The case of the Baltic Sea area. *Baltic Worlds*, 9 (3), 54–67.

Götz, N., Hackmann, J., and Hecker-Stampehl, J., eds., 2006. *Die Ordnung des Raums. Mentale Landkarten in der Ostseeregion*. Berlin: Berliner Wissenschafts-Verlag.

Gould, P., 1966. *On mental maps*. Ann Arbor: University of Michigan.

Gould, P., 1975a. *People in information space: the mental maps and information surfaces of Sweden*. Lund: Lund Studies in Geography.

Gould, P., 1975b. Acquiring spatial information. *Economic Geography*, 51 (2), 87–99.

Gould, P. and White, R., 1974. *Mental maps*. Harmondsworth: Pelican.

Grassland, C., 2012. *Eurobroadmap. Project final report*. European commission. http://cordis.europa.eu/docs/results/225260/final1-final-report-23-juillet.pdf.

Hägerstrand, T., 1953. *Innovationsförloppet ur korologisk synpunk* [Innovation diffusion as a spatial process]. Lund: Gleerup.

Hannes, E., *et al.*, 2012. Mental maps and travel behaviour: meanings and models. *Journal of Geographical Systems*, 14 (2), 143–165.

Holmén, J., 2017. Mapping historical consciousness: mental maps of time and space among secondary school students from ten locations around the Baltic and Mediterranean Seas. *Journal of Autonomy and Security Studies*, 1 (1), 46–75.

Jackson, T.N., 2009. Ways on the 'mental maps' of medieval Scandinavians. *In*: W. Heizmann, K. Böldl, and H. Beck, eds. *Analecta septentrionalia: Beiträge zur nordgermanischen Kultur- und Literaturgeschichte*. Berlin: de Gruyter, 211–220.

Klinge, M., 1995. *Itämeren maailma* [The Baltic Sea world]. Helsinki: Otava.

Lynch, K., 1970. *The image of the city*. Cambridge: MIT Press.

Madaleno, I.M., 2010. How do remote southern hemisphere residents perceive the world? Mental maps drawn by East Timorese and Mozambican islanders. *Scottish Geographical Journal*, 126 (2), 112–136.

Rajaram, P.K. and Grundy-Warr, C., 2007. Introduction. *In*: P.K. Rajaram and C. Grundy-Warr, eds. *Borderscapes: Hidden geographies and politics at territory's edge*. Minneapolis: University of Minnesota Press, ix–xl.

Saarinen, T., 1973. Student views of the world. *In*: R.E. Downs and D. Stea, eds. *Image and environment. Cognitive mapping and spatial behavior*. London: Edward Arnold, 148–161.

Sassen, S., 2002. Towards post-national and denationalized citizenship. *In*: E.F. Isin and B.S. Turner, eds. *Handbook of citizenship studies*. London: Sage, 277–291.

Thill, J.-C. and Sui, D.Z., 1993. Mental maps and fuzziness in space preferences. *The Professional Geographer*, 45 (3), 264–276.

Thomas, M., ed., 2011. *The French colonial mind. Vol. 1. Mental maps of empire and colonial encounters*. Lincoln: University of Nebraska Press.

Tournadre, J., 2014. Anthropogenic pressure on the open ocean: the growth of ship traffic revealed by altimeter data analysis. *Geophysical Research Letters*, 41 (22), 7924–7932.

Tuan, Y-F., 1975. Place: an experiential perspective. *Geographical Review*, 65 (2), 151–165.

The mental maps of Italian entrepreneurs: a quali-quantitative approach

Dario Musolino

ABSTRACT
Geographers interested in how entrepreneurs perceive locational environments have studied their mental maps in several European countries, within the theoretical framework provided by behavioral approach. Such studies have typically employed quantitative techniques, but qualitative studies are relatively new to this line of research. In this article, I examine the mental maps of entrepreneurs in Italy by using a mixture of qualitative and quantitative methods. I present and discuss the qualitative outcomes of this research, focusing in particular on the explanatory location factors and the key influences on the mental maps of entrepreneurs. What emerges is the realization that entrepreneurs are far from being fully rational economic actors, who exploit optimally all information and who are driven only by objective considerations. Rather, their views are also affected by subjective factors, individual's own insights, commonplaces, stereotypes, and prejudices, particularly with reference to the southern regions of Italy (Mezzogiorno), and of other peripheral areas.

Introduction

Mental maps are a long-standing topic of research that has been widely explored and studied since the 1970s, within the theoretical framework provided by behavioral geography (Gould 1966; Gould and White 1974; Tuan 1975; Saarinen 1995), in parallel with key concepts like "perception" and "spatial cognition" (Dietvorst *et al.* 1984; Stern and Krakover 1993), and "image" (Van den Bosch 1977; Pellenbarg 1985). The concept of mental maps at first glance is similar to the other concepts, since they all refer to the subjective perspective of people on places, and are all influenced by aspirations, attitudes, experiences, feelings, ideas, and opinions. However, the concepts are actually quite different.

Perception concerns the way how a phenomenon, an object, or a place, is perceived in a psychological sense by each individual. This perception is then filtered and stored as knowledge, according to "previous cognitive structures in the brain" (Golledge and Stimson 1987, p. 38).[1] In this respect, perception is a subjective part of cognition. This, therefore, should be treated as both objective and subjective knowledge. Consequently, spatial cognition should be understood as the "objective and subjective knowledge of spatial structures, units and relations" (Meester 2004, p. 29).

Place image, meanwhile, concerns the "feelings and impressions about a place" (Spencer and Dixon 1983) by individuals or by a group of people. In this second case, the same feelings and impressions about a place are shared by many individuals (Pellenbarg 1985), and become a collective impression of that place. Such impressions can considerably affect and influence the way people behave. According to the marketing literature, we can distinguish between projected and received place image (Kotler *et al.* 1993). The former regards the ideas and impressions of a place available for people's consideration; the latter is the result of the interaction between the projected messages and people's own needs, motivations, preferences, and other personal characteristics (Angelis and Dimaki 2011).

The term "mental mapping" has taken on different forms and meanings (Meester 2004). The first and most intuitive concept of a mental map is as the image of a place that exists in someone's mind. Second, it can be used to refer to a sketch map drawn by an individual to represent their spatial understanding of an area (Saarinen 1995).[2] Third, it can refer to a knowledge map, which consists of the cartographic representation of spatial knowledge about "objective" conditions, such as the existence of spatial units, or of spatial conditions such as geophysical characteristics or infrastructure (Dietvorst *et al.* 1984). Finally, mental maps can refer to cartographic representations of attitudes and preferences that people hold about places, or preference maps (Tuan 1975; Meester 2004, p. 31).

This last concept of mental map differs from the other relevant key concepts used in behavioral geography. It concerns not only the perceptions and impressions of one place, but the way that several places are simultaneously perceived and evaluated, rated, and ranked by an individual or group of people. In this study, I used and applied this last concept of mental map, the preference map.

Using this conceptualization, a group of economic geographers studied the mental maps of entrepreneurs focusing on their evaluation of locational environments in European countries, such as the Netherlands, Germany, and Czech Republic (Holvoet 1981; Meester 2004; Meester and Pellenbarg 2006; Spilkova 2007; Pellenbarg 2012). These studies, all of which follow the same methodological approach, were based on a sample survey of manufacturing and services entrepreneurs whose key element was a map of the country under investigation depicting the locations to be rated. They aimed, in particular, to describe and explain the shape, patterns, and explanatory factors of the mental maps of entrepreneurs, mostly using quantitative methods such as principal components analysis and regression analysis. The studies conducted, for example, on the mental maps of Dutch and German entrepreneurs highlighted the explanatory importance of several objective and subjective factors. Objective location factors included characteristics such as geographical centrality, high level of accessibility, and the presence of agglomeration effects; areas displaying such characteristics were typically in the "peak area" of the entrepreneurs' mental maps. Subjective characteristics also played an important role, most notably locational self-preference ("the best place is the one where I already am") in shaping their mental maps. Thus, there is an attitude and special attachments towards particular places that are independent of that place's objective characteristics (Meester 2004). Spilkova's research on foreign entrepreneurs in the Czech Republic similarly observed the importance of location factors like geographical location, as well as other factors such as education level and qualifications of the labor force (Spilkova 2007).

Whether the study concerned objective or subjective characteristics, however, the existing research in this area has relied mainly on quantitative research techniques; few have employed qualitative research techniques, in order to explore new characteristics and explanatory factors (with notable exceptions being Meester 2004 and Spilkova 2008). In this study, thus, I examine the mental maps of Italian entrepreneurs through the use of qualitative research methods as well, in order to acquire a deeper understanding of the basic reasons, the underlying motives (in particular, the subjective motives) that can help explain the mental maps of entrepreneurs. To achieve this, as a follow-up to the respondents' ranking of places, I asked two open questions regarding their reasons for rating the best and the worst locational environments.

As a means of further examining and interpreting the results, I followed up with direct semi-structured interviews with experts in attraction of investments and related fields, in order to have a point of view different from the one expressed by entrepreneurs, with the aim of casting further light on the key explanatory factors in the shaping of the entrepreneurs' mental maps.

The article, after presenting the main outcomes of the survey conducted in Italy, focuses on the results of these qualitative analyses.[3] It, therefore, aims at enriching and deepening our understanding of the mental maps of entrepreneurs, not only through their views, but also through the views of other individuals who are well-informed about the subject as well as the behavior and way of thinking of entrepreneurs.

Such emphasis on the subjective perspective of economic actors like the entrepreneurs might be counter-intuitive, if we follow the neoclassic approach to locational theory, which assumes that behavior is rational, objective, economic, and predictable. The ultimate goal of this work then is to investigate and discover factors influencing locational behavior that cannot be effectively ascertained by quantitative statistical techniques.

Such subjective approaches, while uncommon in research on firms' location, have been successfully employed by cultural geographers in studying mental maps in a variety of contexts, such as the influence of historical and political heritage on the mental maps of people living in segregated areas (Smiley 2013), the role of hegemonic ideology and culture on the representation of space (Sletto 2002), and the effects of stereotypes and commonplaces on the regions they depict (Vadjunec *et al.* 2011).

The structure of this paper is as follows. The first section sketches the methodological approach used for surveying the mental maps and shows the results in a series of choropleth maps mapping the average rating of locations. The second section shows and discusses the outcomes of the content analysis of responses to the open questions. The third section presents the results of the thematic analysis of the direct interviews. Finally, the last section draws some concluding remarks.

The mental map of Italian entrepreneurs: methodology and key results

The research on the mental maps of Italian entrepreneurs was based on a web survey of entrepreneurs in leading firms with more than 20 employees, belonging to a range of industrial and services sectors and branches.[4] The electronic questionnaire was quite simple and relatively quick to fill in: the key element was an interactive map of Italy showing the spatial units to be rated (administrative regions and provinces).[5] The respondents had to evaluate each region and province as a possible location for their hypothetic investments on a five-point ordinal scale ("very unfavorable"; "unfavorable"; "neutral"; "favorable"; "very favorable"). The question, accompanying the map, asked respondents to evaluate regions and provinces. In Italian, it reads:

> Supponga che, per qualsiasi ragione, debba cambiare localizzazione alla sua impresa (o a una unità locale della Sua impresa) all"interno del territorio del nostro paese. Sulla base di questa ipotesi, che valutazione da', come possibile localizzazione, ad ognuna delle aree indicate nella mappa allegata?

In English:

> Suppose that, for any possible reason, you have to change the location of your firm (or of one the units) within the Italian territory. Given this hypothesis, how do you evaluate each of the areas indicated on the map as possibile new locations of your firm?

Respondents evaluated the two geographical levels following a stepwise mechanism: they had to first rate regions and then, eventually/optionally, provinces. This is why it is also important to take both maps (regional and provincial) into account when analyzing the results of the web survey, since they were not rated fully independently from each other. I contacted all entrepreneurs satisfying the criteria (about 10,000) to participate in the survey between January 2010 and July 2011. The return was 645, of whom 225 properly filled out the questionnaire, making them usable for analysis.

The main results of the web survey, the mental maps at regional and provincial scale respectively of the entrepreneurs participating in the survey, can be seen here in two thematic maps (Figures 1 and 2). At first glance, it is evident that two spatial patterns characterize both scales of mental maps: the North–South divide, which is a historically rooted and persistent characteristic of Italian economic geography,[6] and the center–periphery dichotomy. Indeed, the spatial hierarchy from north to south is evident. In addition, a second pattern, even clearer from the map at provincial scale, is the center–periphery dichotomy, where the center is the Padana region (Lombardy, Veneto, and Emilia-Romagna), and with the area around Milan constituting the core.[7]

A number of important explanatory points emerged from the data. The first point is that these two patterns are shared by all kind of entrepreneurs. In fact, when dividing entrepreneurs by type (by economic sectors, firm size, sex, age and education level), their mental maps do not show any significant differences, except for the ratings of a few areas. Second, it came to light that some location factors usually taken into consideration by the location theory and by several empirical studies – i.e. accessibility and agglomeration economies – proved to be rather important in explaining the high rating of the northern regions and provinces (in particular, the ones located in the Padana region), and the low ratings in southern regions. However, such factors were not enough to entirely explain the results. Thus, I turned next to a content analysis of the entrepreneurs' qualitative explanations of their responses, as well as a thematic analysis of the direct interviews.

Motives for place ratings: evidence from content analysis

Part of the electronic questionnaire consisted of two open questions, inserted after the interactive map, where I asked respondents to explain the rating they gave to two of the best-rated provinces, and two of the worst-rated provinces. The two open questions asked as follows:

(1) "We have noticed that the province of _______ and the province of _______ are among the provinces that you have evaluated as the best hypothetical locations for your firm. Can you shortly explain the reasons why you gave this evaluation, for each of them?"

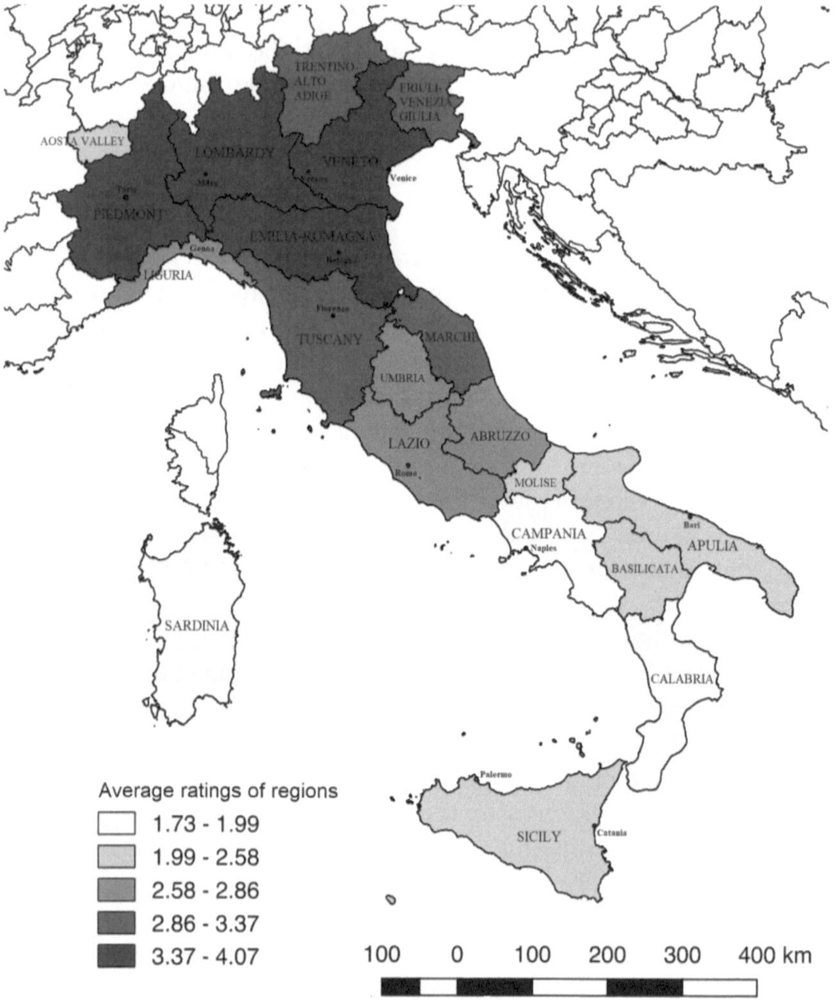

Figure 1. Average ratings of Italian Nuts2 regions, as places where to hypothetically locate investments (Source: elaboration on data from questionnaire survey, 2011; 225 usable questionnaires; 5-point ordinal scale: 1: "very unfavorable"; 2: "unfavorable"; 3: "neutral"; 4: "favorable"; 5: "very favorable"; arithmetic mean: 2.82), and main Italian cities (municipalities over 250.000 inhabitants, as from Istat).

(2) "We ask you the same question as concerns some provinces which, on the contrary, you have evaluated as the worst hypothetical locations for your firm (province of ______, and province of _______). Can you shortly explain the reasons why you gave this evaluation, for each of them?"

The software randomly chose these four provinces from among the respondents' answers. I analyzed the responses to the two open questions

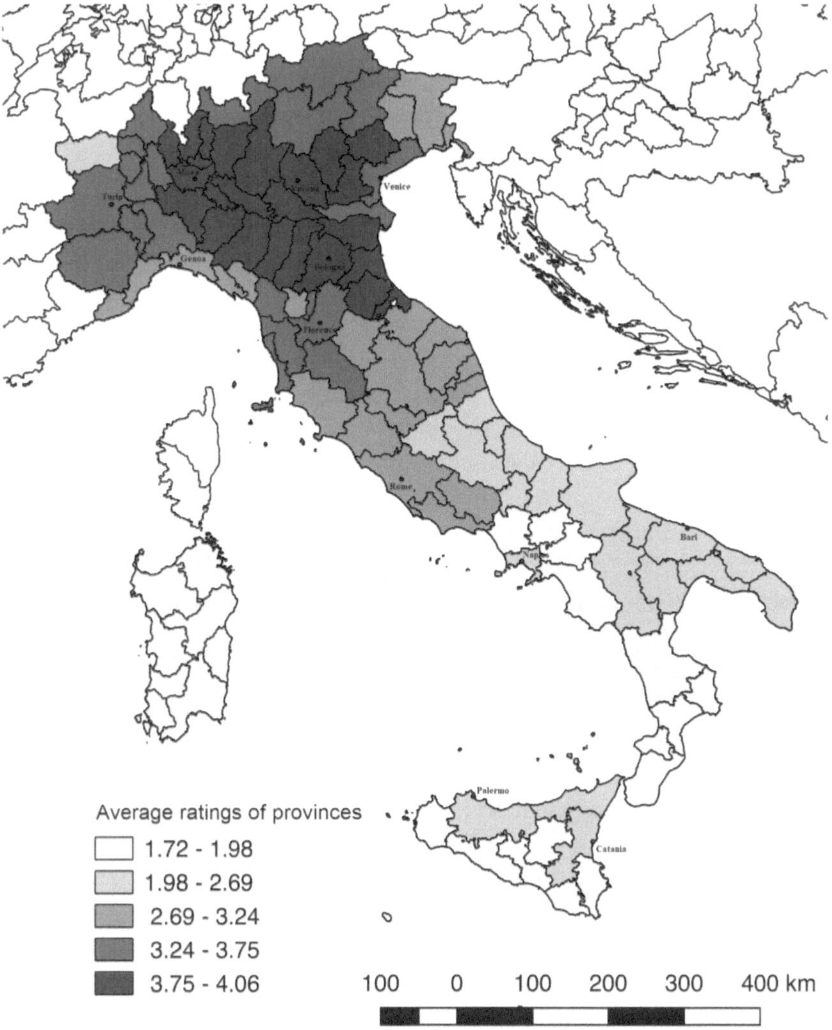

Figure 2. Average ratings of Italian Nuts3 provinces, as places where to hypothetically locate investments (Source: elaboration on data from questionnaire survey, 2011; 225 usable questionnaires; 5-point ordinal scale: 1: "very unfavorable"; 2: "unfavorable"; 3: "neutral"; 4: "favorable"; 5: "very favorable"; arithmetic mean: 2.92).

using content analysis, and in particular the technique of "category counts" (Stone *et al.* 1966; Rositi 1988). The "category counts" technique identifies some keywords in the text of each answer, counts, and defines a certain number of thematic categories based on them. Thematic categories were aggregated following an inductive approach (they emerged from the examination and the comparison of the data). However, I have taken into account the previous studies using qualitative data coming from direct interviews with entrepreneurs. For example, in the study about the Netherlands (Meester 2004, pp. 115–121), the author identified 12 categories following a inductive aggregation approach (relative location; infrastructure; accessibility; ties to one's own region; mentality of the population; living environment; labor market; competition issues; agglomeration effects; government; characteristics of the premises; and property expenses).

One hundred and twenty respondents answered the two open questions, but only 107 gave complete and valid responses. The respondents used 279 keywords regarding positive location factors, and 254 keywords regarding negative location factors. The aggregation of these keywords resulted in 12 categories.

The key location factors that more frequently recurs for both the best-marked and the worst-marked provinces (Figure 3) are transport services and logistics (18.6% of the total number of the mentions for the best-marked provinces, and 26.4% for the worst ones), together with the proximity to markets and suppliers (18.3% and 16.1%), and the geographical location (14.7% and 14.6%). In the case of the worst-marked provinces, located in Southern Italy, the importance of these three location factors in the explaining the mental of entrepreneurs is even bigger than for the northern areas. These three factors clearly refer to the question of the accessibility and the agglomeration economies: they are, therefore, the key issues that explain why entrepreneurs perceive some locations in Northern Italy to be so advantageous (they are supposed to be enough endowed with transport infrastructures and services, and to have a sufficient level of agglomeration economies), and why

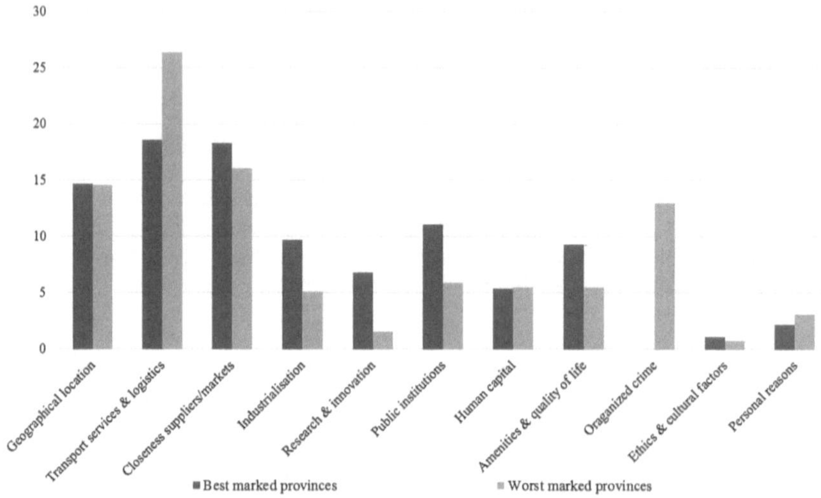

Figure 3. Location factors mentioned for the best- and the worst-marked Italian Nuts3 provinces (elaboration on data from questionnaire survey, 2011; 225 usable questionnaires; percentage values).

they consider some Southern areas so neglected as potential locational environments (a neglect that Svimez (2013) termed "industrial desertification").

For the worst-marked provinces in Southern Italy, security concerns related to the presence of organized crime also were a significant factor (13%), related to the presence of the organized crime. These organizations are notoriously strongly rooted in some areas in Southern Italy, where, to a certain extent, they have the power to influence and control local economic activities (see, e.g. Daniele and Marani 2011).

Although less important, the role of other factors emerges as well, particularly with the best-marked provinces. Accounting for at least 5% of the total number of positive mentions, in order of importance, they are: the presence of efficient public institutions and of public policies that foster firm location; the "industrial atmosphere" and the development of the industrial system (a factor that can intuitively be connected to the agglomeration economies); the presence of amenities and the quality of life that contribute to create a favorable environment for the firms and for their employees; the attitude to research and innovation that can benefit product and process innovation of the firms located there; and the availability of human capital.

In the case of the northern areas, there is a wider diversification of the factors that have importance in shaping the mental maps of entrepreneurs. In the mind of entrepreneurs each northern locational environment tends to have its own peculiarities, its own "specializations" or "assets" that make it attractive. On the other hand, in the case of the (negatively rated) southern areas, fewer location factors strongly catch the attention of entrepreneurs, and they tend to be negative: they, therefore, negatively affect, and simplify, the image of all of entire Southern Italy.

The underlying location factors: evidence from the thematic analysis

After the first results of the survey became available, I conducted nine direct semi-structured interviews with key informants, experts in attraction of investment and related fields, between July 2013 and February 2014. The aim of these interviews was to investigate the opinion of a range of experts on the findings of the web questionnaire survey, in particular about the average ratings of regions and provinces, and about the main patterns that emerged from them. I asked them, for example, whether they shared the evaluation given by entrepreneurs to regions and provinces. Then, I asked if, according to them, there was something "wrong" in the perceptions of entrepreneurs (patterns different from those that they would have expected), and how they could explain the discrepancy. In the process, we discussed the key arguments, the key subjective and objective factors that could explain and determine the territorial patterns visible in the maps.

The ultimate objective of the interviews was to illuminate underlying motives and driving forces that might explain the perceptions of entrepreneurs. For these interviews, we targeted consultants, representatives of local, regional and national bodies devoted to implementing policies to attract of direct investments, representatives of manufacturers' associations, experts on the issue of firms' locational choices, and institutional investors. I obtained these contacts from a variety of sources, and tried to maintain a degree of representativeness in terms of macro-geographical location (northern, central, and southern). The main objective of this research strategy was to learn the opinion of a range of key informants on the question of the perceived attractiveness of places, but from a point of view typically different from that of entrepreneurs, from the perspective of people with different backgrounds and experiences. Of course, needless to say, even experts can have their own biases and their mental maps, which might influence their responses. The aim clearly of the interviews was not to search for a more "correct" point of view, rather it was to ask for an opinion different from the one expressed by entrepreneurs. Table 1 describes in details the characteristics of each of the interviewees.

I conducted the interviews on the basis of an outline, including the basic figures and tables synthesizing the key results of the questionnaire survey, which was sent in advance to the person who agreed to be interviewed. Interviews were conducted in Italian, and then translated into English. After the consent of the interviewees, I analyzed the transcripts of the direct interviews using thematic analysis. This technique serves researchers who are using qualitative data, especially verbal expressions coming from open interviews, in "identifying, analyzing, and reporting patterns (themes and sub-themes) within data" (Braun and Clarke 2006, p. 6). The identification of the relevant key themes depends on the judgement of the researcher, "in terms of whether it captures something important in relation to the overall research question" (Braun and Clarke 2006, p. 10). Therefore, it is not associated with any quantifiable measures.

In the examination that follows, I identified recurrent themes and subthemes, not necessarily linked to the outline of the interviews. I frequently used direct quotations in the text in order to provide clear evidence about the relevance and the prevalence of a theme, as was done in other studies using thematic analysis (see, for example, Ellis and Kitzinger 2002; Kitzinger and Wilmott 2002; Delaney *et al.* 2011).

Accessibility and agglomeration: key factors for the shape of the mental maps

The interviewees definitely support the impression that Central-Northern Italy and Southern Italy are divided in terms of attractiveness for direct

Table 1. Characteristics of the interviewees.

N	Kind of institution	Place	Position	Sex and age group
1	Agency for the promotion of the International activities of the Chamber of Commerce of a big Italian city	Milan (Lombardy)	Global development manager	Male (30–39)
2	Management company of a closed-end investment fund	Milan (Lombardy)	Partner, sitting on the board of several medium-sized manufacturing Italian companies	Male (40–49)
3	Italian association of the firms belonging to one of the main manufacturing industries	Milan (Lombardy)	Responsible for researches and statistical analyses	Male (50–59)
4	Central government agency for inward investment promotion and enterprise development	Rome (Lazio)	Director	Male (50–59)
5	Integrated firm for audit, legal, management, and tax consulting	Milan (Lombardy)	Managing Partner (Consultant for German firms investing in Italy)	Male (50–59)
6	Agency for the promotion of international activities of one of the Northern regional governments	Turin (Piedmont)	Top manager	Male (30–39)
7	Department for inward investment promotion policies of one of the Southern regional governments	Palermo (Sicily)	Director	Male (40–49)
8	Medium-sized company belonging to the food sector, involved in regional policies for inward investment promotion	Ragusa (Sicily)	Managing Director	Male (40–49)
9	Consultant for agencies for the attraction of foreign direct investments, and for local, regional and national governments in several countries, including Italy	Paris (France)	Individual consultant	Male (40–49)

investment, so they confirm the existence of the main pattern (north–south divide) emerging from the analysis of the entrepreneurs' data. However, they suggest the relevance of other spatial patterns, such as the center–periphery dichotomy and the prominence of the Padana region. According to them, the basic explanatory elements for the dichotomic patterns of the mental maps of entrepreneurs can be found in two interrelated issues: accessibility, which includes the geographical centrality, and agglomeration economies.

What they point out, in fact, is that the high concentration, density and proximity of economic activities, firms, services, and people[8] in Northern Italy, and in particular in the Padana region, and the considerable endowment of infrastructure available, in particular for transport,[9] create economic advantages which attract and favor the location of new businesses and the growth of existing ones, coherent with a process of circular cumulative causation (Myrdal 1957). Such a phenomenon made this macro-region historically the largest and most important business environment in Italy, far ahead other macro-regions in terms of locational advantages, in particular as compared to southern Italy. Furthermore, the interviewees suggest that the central geographical location with respect to the European continent significantly contributes in making northern Italy, especially the Padana region, an ideal place to start a new business. Northern Italy is geographically closer to many European markets than to Southern Italy, which is peripheral in the geo-economic European context (and less accessible, since its transport networks and services are not adequate and efficient). As stated by interviewee 4:

> Such an agglomeration is for sure an outcome of history. There is such an agglomeration because time after time it grew and it consolidated ... In that region there are very important rail and air infrastructural nodes, so if a firm has to choose where to locate ... such mechanism is a virtuous circle: when all people want to be there, those who have all their customers there, will locate there, or those who have all their suppliers there, will then locate there, and then other customers and suppliers will locate there, new infrastructure will be built, and so on, and proximity effects, network effects will exert their impact. ... Maybe such location factors are obvious, banal, but ultimately they are decisive ...

Within the Padana region, many of interviewees highlight the role played by the area of Milan, which, thanks to its advantages in terms of geographical centrality, transport nodality, and agglomeration economies in particular as concerns service sectors (OECD 2006), tops the perception of Italian economic geography by entrepreneurs, and polarizes further their perception. According to interviewee 1:

> Surely, compared with other locations in Italy, Milan is seen as the center of business. It is central, rich in services, internationally known and open, highly oriented to commercial activities ... Italian firms think that it is the center of business, the base, the headquarters of their activities ...

As pointed out by interviewee 9:

> Not only about the North–South divide, we should also talk about the Milan–'-other than Milan parts of Italy' divide … Milan polarizes the map in particular because of its attractiveness for investments,[10] especially in the field of sales marketing. If there were more investments in manufacturing [elsewhere in the country], the map of Italy would be less unbalanced …

The key role of Milan, and of the Padana Region, can be even better understood if we compare it with the other main urban agglomeration in Italy, the capital city, which is the political and institutional center. For interviewee 1:

> In the view of a foreign firm, Rome is a base for lobbying, to have relations with public actors, while other reasons are less important. Milan, instead, is the base to do business at the international level, where English is widely spoken.

The paradox of the northern peripheral areas: developed but "unseen" by entrepreneurs

The consequence of such location factors in the minds of entrepreneurs, and of the tendency to dichotomize locational environments, is the ranking gap observed in Northern Italy between the central plain regions (Lombardy, Piedmont, and Veneto) and the surrounding coastal and mountain regions, such as Friuli, Aosta Valley, and Liguria. According to the interviewees, this is one of the most interesting, and somehow astonishing results emerging from the survey, because, in terms of real socio-economic indicators, such as GDP per capita and the employment rate, these regions actually do not lag much behind the core Northern regions (see, e.g. Musolino 2015, ch. 2).

It is true that these areas do not enjoy the support of driving forces such as agglomeration economies, high accessibility, and geographical centrality. For example, interviewee 9 points out the marginality of Liguria in terms of physical geography and of infrastructure. Interviewee 2, referring in particular to Friuli and Liguria, underlines that these regions suffer from the lack of big firms which typically are leaders of big value chains which locate and agglomerate close to it:

> … all big firms that were there either failed and were dismantled … such dynamics causes the death not only of the related supply chain, providing goods and services, but also of all the spin-off firms born from these big firms, set up by managers, employees, who left the leading firm … when the "mother" fails, all the other small firms which strongly depend on her, fail too …

At the same time, it is evident that these Northern "peripheral" regions, such as Trentino-Alto Adige and Aosta Valley, have important specific locational advantages that are "unseen" by most of the entrepreneurs. For example, according to interviewee 5, who focuses on the case of Trentino- Alto Adige, and on its particular culture and language:

> Some German firms, and also Austrian firms, start in this region, because of the language, and then from there they try to spread all over the Italian territory ... In Trentino-Alto Adige they can write the certificate of incorporation in Italian, and also they can start to explore the Italian market with people who speak German. And so, if things go well, they also go into other regions, for example, Veneto ... it is a kind of linguistic and cultural advantage.

Interviewee 4 added that, even if Trentino is a small region where necessarily the mass and size effects are less strong, there are significant areas of specialization, "some highly competitive niches".[11] Interviewee 6, talking about Aosta Valley, finds its average rating not understandable, given the great effort of its regional government in supporting firms:

> Aosta Valley is seen as a very attractive region, a very "friendly" region for firms ... Many Piedmont firms who move there can have several incentives, many benefits ... Aosta Valley regional government, for example, supported some entrepreneurs in finding the areas where to build their new plant, getting very low land prices ... so I am surprised that such aspects do not emerge from the survey. Maybe the issue is that it is a small region.

The stereotypical image of the Mezzogiorno and the role of the Mafia

If accessibility and agglomeration economies explain much of the north–south divide, Southern Italy also suffers from the presence of an anomalous factor, a specific motif that affects the perception of entrepreneurs, as the content analysis of their responses also shows. A number of interviewees highlighted the role played by organized crime in influencing the mental maps of entrepreneurs. According to them, the Mafia more than other factors, creates a prejudice against the Italian Mezzogiorno (and to some degree against all of Italy), that right from the beginning prevents (mostly foreign) investors from taking those regions into consideration.

Organized crime can have a negative influence on the regular economic activities (Asmundo and Lisciandra 2008; Bonaccorsi di Patti 2009; Asso e Trigilia 2011). It generates direct and indirect costs for regular firms, such as extortion and constraints in recruiting workers and in applying for public tender notices (for example, in the construction sector). Moreover, by using their criminal power to protect their own "legal" economic activities, crime organizations discourage other firms from competing with them (their own firms can, therefore operate as monopolies); and, by using illegal financial resources to lower their costs, "legal" firms under their control can be unfairly competitive on the free market. From the initial stage of the locational decision-making process, this key factor induces them to exclude all Southern areas from the range of places where they will consider locating a new plant. As suggested by interviewee 4:

> Many investors tend to say: south of Rome we do not take any place into consideration … the main reasons why they do not value the characteristics of the locational environments south of Rome is the issue of organized crime. This question affects all Southern regions … .

Also interviewee 9 said that:

> The main problem that comes to mind when one thinks of the Mezzogiorno is the question of safety of the plant and of the people. It is hard to think about making investments, in particular highly capital intensive investments, in such places where they do not guarantee security.

Interviewee 4 reported about the case of a multinational company that preferred to insert in its shortlist of the potential locations of its investment in Italy, a central region instead of a Southern region, despite the latter offered much higher tax benefits.

As studies on crime organizations have shown (Fondazione Transcrime 2013), the presence of the Mafia varies in Southern Italy. Whereas there are Southern regions, where they are historically rooted and influential in the local economy (Campania, Calabria and Sicily), in other Southern regions, such as Basilicata, Abruzzo, Sardinia, and Molise, there is a very scarce presence of this phenomenon. On the contrary, in some northern and central regions of the country, mafia organizations are significantly present.[12] Nonetheless, the skeptical attitude towards Southern Italy involves all southern regions and provinces, engendering a stereotyped image of Mezzogiorno. This renders an undifferentiated representation of the South as a macro-region where the fear of organized crime threatens any starting or existing entrepreneurial activity. In fact, the standard deviation of the average ratings of regions and provinces in Southern Italy is much lower than the variance for Central and Northern Italy, although both macro-regions are rather heterogenous in socio-economic terms (Musolino 2015, p. 51).

The "real" Mezzogiorno? A heterogeneous, not flat, space

The reality, according to the interviewees, is different. The economic geography of Southern Italy, as also shown by socio-economic statistics, is quite heterogenous (Svimez 2013, ch. 3.2). For example, a number of interviewees address the situation of Apulia, with some peculiar factors like the straightforward and sectorally focused economic development strategies of the regional government, that have contributed in changing and improving the efficiency of the regional and local bureaucracy. According to interviewee 4,

> In Apulia public institutions have a new approach, more professional, there is a kind of upgrade of all regional government structures … It is a matter of policy. Apulia decided, for example, that renewable energies are a focus sector for

> economic development policies. There were strong and clear decisions in terms of industrial policy (i.e. policy for investment attraction), which have been followed since then. When you do that, investors react positively and follow you.

There is not only the case of Apulia. For example, according to interviewee 7, some areas in Sicily have been able to boost their level of economic development, thanks also to the attraction of exogenous investments in sectors such as tourism, real estate, and agri-food. Local policy makers were able to create an attractive environment by improving contextual factors, (public and private services), such as services related to tourism and cultural heritage, and the health system. In turn, interviewee 4 made reference to Campania, and to its provinces, which are highly specialized in some sectors:

> Its rating should be close to that of Apulia. When we meet foreign investors it frequently happens that, if they take the Mezzogiorno into account, the regions that are almost always included in the shortlist are Apulia and Campania. Instead, I see in this map that Campania is closer to Calabria than to Apulia, notwithstanding its industrial system in all the provinces of Campania, its excellent places … as I see it, this is something explained by a wrong perception. Probably, it depends on the fact that when you asked respondents to rate Campania they immediately thought only about Naples, and about its well-known dramatic problems, such as the waste management crisis or "camorra".

The (not negligible) role of public institutions and human capital

The low level of legal certainty caused by the lack of political stability of the central government, and the low level of efficiency and effectiveness of public administration, are notorious factors that negatively affect the attractiveness of the entire Italian economic system (Siemens-European House Ambrosetti 2007; IPSOS 2008; Cannari and Franco 2010). A number of interviewees are aware of these weaknesses and refer to them. For example, interviewee 1 regarded the issues of legal uncertainty and inertia as the most inhibiting factor for foreign investors, even more than taxation. Interviewee 3 specified that in their sector the critical question was the uncertain application of the law by the bureaucracy, rather than the laws themselves. According to the interviewees, in this respect some aspects differ between the Northern-Central regions and the Southern regions, and can contribute to explain the shape of the mental maps of entrepreneurs, although they do not so frequently mention them, as can be seen in Figure 2. As said by interviewee 8, speaking generally about the quality of administration:

> Italy is a highly bureaucratized country, and we know that, but between North and South there is a difference, for sure. Some things go slower in Southern Italy. In our firm we took one year and 6 months to apply all the relevant rules, all licenses, authorizations, all what we needed. This is the time that bureaucracy takes in Southern Italy. I see that the Northern bureaucracy is

> faster as it concerns some things, and they even exert controls much more frequently.

Interviewee 5 pointed out the issue of trial length as a typical example of North–South differences in terms of the efficiency of public institutions. According to him, the average trial length in Southern Italy was higher. His agency, therefore, informed its customers that in the Mezzogiorno, legal controversies were drawn out, and found it hard to explain why it could take years to hold a juridical hearing.

Another factor relevant in explaining the spatial patterns seen above, differing from what the content analysis would suggest, is the quality of human capital. According to a number of interviewees, in fact, a divide between Northern-Central and Southern regions can be observed in entrepreneurs' perceptions with regard to the human capital factor as well, both in general terms, and with reference to particular skills, competencies, and specific stages of the education process.[13] As pointed out by interviewee 5, not only was the supply of qualified people probably bigger in the North of Italy, but there was also a sort of brain drain caused by people leaving the South for study or work in the North. According to interviewee 2,

> There is a broad perception that we still have a macro-regional divide in Italy as it concerns some basic skills. For example, my idea is that nowadays recruiting someone in Southern Italy who speaks English quite well is more difficult than in Northern Italy.

And interviewee 2, focusing on the mismatch between demand and supply on the Southern labor market, pointed out that a considerable portion of Southern students chose to build skills in less career-oriented fields such as the humanities.

Amenities and quality of life: the positive face of all Italian regions

Lastly, the amenities and the quality of life is a factor that seemingly does not produce a picture that fully coincides with the predominant perception of the entrepreneurs. In this case, according to the interviewees, the map does not show a dichotomic pattern or a core, but this factor simply engenders a flat map, where all places seemingly have the same potential attractiveness.[14] Using the words of interviewee 5, it seems that "Italy has got two faces: one, not so related to business, is absolutely positive, the Dolce Vita, the Belpaese, the delicious food products, the fashion." The quality of life factor creates a positive image, which somehow balances the negative prejudices determined by other factors. That is important especially in the first stage of the decision-making process concerning locational choices, according to interviewee 4:

> That's the problem, of course. In the sense that—thinking with the mind of foreigners—it is one of the reasons why, at least, in the initial stage, they take Italy into consideration. Because I believe that all, and in particular non-European, investors, when they think about Europe, they would not mind being in Italy ... we lose something regarding other aspects, but this factor for sure mitigates the effects of negative prejudices ... just as we have negative prejudices which prevent either the country as a whole or some part of it from being taken into consideration, we have also some factors which, thanks to positive prejudices, favor us, and let us be the first in the shortlist ... the final decision is usually taken on the basis of other pragmatic factors, but other things being equal I remember that a Far Eastern multinational company preferred Florence to Paris.

Beyond building a positive image, amenities can be important for the type of investments where talented, highly qualified human resources are involved, such as headquarters and research and development units. Interviewee 9 highlights:

> It will be always more difficult to bring a smart and intelligent foreign researcher to a place where there is no international school for his children, where the level of quality of life is low, where there are no cultural activities.

Conclusions

The first reflection on the qualitative analyses presented and discussed above is that the mental maps of Italian entrepreneurs are dichotomic. The North–South divide and the center–periphery dichotomy (Padana region versus other areas) are the main spatial pattern observed, as confirmed by the key informants, and they are clearly associated with three other key explanatory dichotomies: high vs. poor accessibility; geographical centrality vs. peripherality; and agglomeration vs. dispersion. These explanatory dichotomies are apparently all interrelated, and have historically fostered a kind of mechanism that reinforces them continuously, as theorized by Myrdal (1957), fully capturing the attention of entrepreneurs.

In addition to these explanatory dichotomies, it is evident that the threshold between the main Italian macro-regions – the "wall in the head" (Maseland 2014, p. 1162) in the minds of the entrepreneurs – is grounded on a prejudicial view, a stereotyped image of Southern Italy caused by the presence of the Mafia. Organized crime is apparently the main anomalous and unique factor that, according to interviewees, contributes to such a stereotype, such "lack of confidence" of entrepreneurs in the Mezzogiorno, which they assumed to be equally insecure in all of its regions and provinces. Such dichotomic patterns of the mental maps of Italian entrepreneurs result in an excessive polarization of the perceived geography of Italy, and in an oversimplification of its explanation in terms of location factors, creating a clear inconsistency between the perception and the reality of its economic geography.[13]

Such oversimplification can be seen, for example, in the relevance – pointed out by the experts – of the explanatory role played by public institutions and by human capital. Such factors were rarely mentioned by the entrepreneurs in their responses to the open questions, particularly regarding southern areas. Similarly, the peripheral areas in Northern Italy appear to be seen as unfavorable places from the point of view of entrepreneurs, despite their clear economic dynamism and their locational advantages, such as the linguistic advantage of Trentino-Alto Adige in relation to German business, and the policies of Val d'Aosta for attracting investments. Moreover, the presence of organized crime in some Southern areas apparently overshadows other regions and places in Southern Italy, such as Apulia and some areas in Sicily, in Campania, and Sardinia, partially hiding their own specific locational advantages.

The results of these analyses call, therefore, for further investigations, focusing first of all on the subjective factors, and secondly also on the role on the intermediate agents (media), which reasonably have an effect on the perception of people, as some studies highlighted, for example with concerns Italy and the role played by the Mafia on the voting behavior (Mastrorocco and Minale 2016). This also means that necessarily future research on these topics should open and involve other social science fields, such as sociology, anthropology, and psychology, and other fields in the geographical sciences, like cultural geography.

What emerges finally from the qualitative data collected is that the mental maps of Italian entrepreneurs are far from outcomes of the view of fully rational economic actors, optimizers who exploit systematically all information and follow only objective considerations. Rather, they are affected by subjective factors, "knowledge gaps", individual insights, commonplaces, stereotypes, and prejudices, such as the ones that seemingly affect their perceptions of the southern regions of Italy (Mezzogiorno), and the peripheral areas of the North. The mental maps of entrepreneurs are affected by both subjective and objective factors that tend to make them dichotomic and oversimplified in their structure, hiding potential locational (dis)advantages of Italy's wide and diverse locational environments, and therefore, trivializing the variety and complexity of the economic, social, and cultural geography of Italy.

Notes

1. Perception, in more basic terms, regards the sensory input given to an individual by a phenomenon, an object, or a place, while cognition refers to the information, and the knowledge acquired and used by each individual. According to Stern and Krakover, perception concerns the "usually immediate apprehension of environmental information by our senses, while cognition refers to the way this information, once received, is organized in the brain" (1993, p. 131).

2. For a recent application of the concept of sketch maps, see the study conducted within the project "EuroBroadMap – Europe seen from abroad" aimed at analysing the different visions of students concerning Europe in the world (Didelon *et al.* 2011, Grasland and Beauguitte 2011).
3. The arguments used in this article are based on Musolino (2015).
4. Entrepreneurs had to satisfy three basic criteria, already applied for the other country-level case studies (Meester 2004): (1) being capable to make a well-founded judgment on the locational environments in the study area; (2) having an interest, even hypothetic, to evaluate an alternative location; (3) having the power to take decisions about the location of the plants (that is to say, who can decide about the location of the firm by themselves). The use of these criteria limited the range of sectors and branches part of the research population. For example, firms belonging to sectors with a strong locational constraint, such as activities bound to the land (mining, agriculture, etc.), were kept out, as they do not have any interest in evaluating alternative and different locations, and therefore they do not satisfy the second criteria.
5. These divisions correspond respectively to the NUTS 2 and NUTS 3 levels in Italy, as defined in the Classification of Territorial Units for Statistics by the European Union, the hierarchical system for dividing up the economic territory of the EU [http://ec.europa.eu/eurostat/statistics-explained/index.php/Glossary:Nomenclature_of_territorial_units_for_statistics_(NUTS)].
6. See for example Svimez 2011, chapter XI.
7. For relevant studies about the emergence of the Padana region, see, e.g. Bramanti *et al.* 1992, Turri 2000, OECD 2006.
8. The Padana region in 2015 accounted for 48% of the total GDP and the total value added in Italy (dati.istat.it), and for 43% of firms. The four regions approximately composing Padana region (Piedmont, Lombardy, Veneto, Emilia-Romagna) were inhabited in 2016 by about 23.5 million people, about 39% of the total Italian population (dati.istat.it).
9. According to the latest data about accessibility at the Nuts3 level in European union (Spiekermann and Wegener 2014), the level of accessibility in the regions in the Padana region is usually over the EU average, and in some cases (for example, Milan, Bologna) it is about 50% higher.
10. Milan accounts for 28% of all foreign-owned firms located in Italy. In terms of employees, it accounts for 33% of all employees in foreign-owned located in Italy (Mutinelli 2017).
11. See for example mechatronics, agri-food (apples, grapes, etc.), building (green building) and the furniture sector.
12. According to Fondazione Transcrime (2013), who defined a synthetic index for measuring the presence of Mafia organizations in Italy at the territorial scale in the period 2000–2011, there are relevant differences among Southern regions. While in Campania, Calabria and Sicily, this index measures 61, 32 and 42; in Basilicata, Abbruzzo, Sardinia and Molise it measures respectively 5, 0.7, 7 and 0.3. As concerns other regions, for example in Lazio it is 17, in Ligura it is 10, and in Piedmont, 6.
13. According to Istat data (dati.istat.it, 2015), in Northern regions college graduates account for 13% of the population older than 15, while in Southern regions, the figure is 11.5%.
14. Many of the cultural and other assets often attributed to Italy as a whole can be found readily throughout the country. As a simple example, properties

inscribed on the World Heritage List in Italy (http://whc.unesco.org/en/statesparties/it), which total 53, are relatively evenly distributed throughout the country in all regions.

15. Such inconsistencies are in fitting with other findings of the gaps between perceptions and reality of regional differences in economic indicators such as GDP per capita, where the perceived North–South gap" is wider than the "real North–South gap." (Musolino 2015, chapter 6).

Acknowledgments

I would like to thank the anonymous reviewers, for their revisions and their advice. I am grateful to Norbert Götz and Janne Holmén, Guest Editors of this special issue, and to Steven Schnell, Editor of the *Journal of Cultural Geography*, whose comments and suggestions much improved this paper as well. Lastly, I thank Niccolò Pieri, who supported me in making the two choropleth maps of Italy (Figures 1 and 2).

Disclosure statement

No potential conflict of interest was reported by the author.

ORCID

Dario Musolino http://orcid.org/0000-0002-8245-0798

References

Angelis, V. and Dimaki K., 2011. A region's basic image as a measure of its attractiveness. *International Journal of Business and Economic Sciences Applied Research*, 4 (2), 7–33.

Asmundo, A. and Lisciandra, M., 2008. The cost of protection racket in Sicily. *Global Crime*, 9 (3), 221–240.

Asso, P.F. and Trigilia, C., 2011. Mafie ed economie locali. Obiettivi, risultati e interrogativi di una ricerca. *In:* R. Sciarrone, eds. *Alleanze nell'ombra. Mafie ed economie locali in Sicilia e nel Mezzogiorno*. Roma: Donzelli Editore, 13–36.

Bonaccorsi di Patti, E., 2009. Weak institutions and credit availability: the impact of crime on bank loans. *Bank of Italy Occasional Paper*, p. 52.

Bramanti, A., *et al.*, 1992. *La Padania, una regione italiana in Europa*. Torino: Edizioni della Fondazione G. Agnelli.

Braun, V. and Clarke, V., 2006. Using thematic analysis in psychology. *Qualitative Research in Psychology*, 3 (2), 77–101.

Cannari, L. and Franco, D., 2010. Il Mezzogiorno e la politica economica dell'Italia. *Bank of Italy Workshops and Conferences*, 4, 105–127.

Daniele, V. and Marani, U., 2011. Organized crime, the quality of local institutions and FDI in Italy: A panel data analysis. *European Journal of Political Economy*, 27 (1), 132–142.

Delaney, L., Egan, M., and O'Connell, N., 2011. The experience of unemployment in Ireland: a thematic analysis. *UCD Geary Institute Discussion Paper Series*, August.

Didelon, C., *et al.*, 2011. A world of interstices: A fuzzy logic approach to the analysis of interpretative maps. *The Cartographic Journal*, 48, 100–107.

Dietvorst, A.G.J., *et al.*, 1984. *Algemene sociale geografie: ontwikkelingslijnen en standpunten*. Romen: Weesp.

Ellis, S.J. and Kitzinger, C., 2002. Denying equality: An analysis of arguments against lowering the age of consent for sex between men. *Journal of Community & Applied Social Psychology*, 12, 167–180.

Fondazione Transcrime, 2013. Dove operano le mafie in Italia. *In: Progetto PON Sicurezza 2007–2013. Gli investimenti delle mafie*. Available from: http://www.transcrime.it/pubblicazioni/progetto-pon-sicurezza-2007-2013/ [Accessed 30 October 2017].

Golledge, R.G. and Stimson R.J., 1987. *Analytical behavioural geography*. London: Croom Helm.

Gould, P.R., 1966. On mental maps. *Michigan inter-university community of mathematical geographers (Discussion paper 9)*. Ann Arbor. *Reprint in:* R.M. Downs and D. Stea, eds. 1973. *Image and environment: Cognitive mapping and spatial behavior*. Chicago: Aldine, 182–220.

Gould, P. and White, R., 1974. *Mental maps*. Harmondsworth: Penguin.

Grasland, C. and Beauguitte, L., 2011. *Modelling attractiveness of global places. A worldwide survey on 9000 undergraduate students. Paper presented at ERSA 50th Congress*, August 19–23 Jönköping, Sweden.

Holvoet, M., 1981. Localisation industrielle en Belgique: Cartes mentales et préférences spatiales d'un groupe de futurs responsables économiques. *Revue Belge de Géographie*, 105, 41–57.

IPSOS, 2008. *Attraction Italy. The opinions of the managers*. Milan. Available from http://slideplayer.it/slide/1791796/ [Accessed 30 October 2017].

Kitzinger, C. and Willmott, J., 2002. The thief of womanhood: women's experience of polycystic ovarian syndrome. *Social Science & Medicine*, 54, 349–361.

Kotler, P., Haider, D.H., and Rein, I., 1993. *Marketing places*. New York: The Free Press.

Maseland, R., 2014. Does Germany have an East–West problem? Regional growth patterns in Germany since reunification. *Regional Studies*, 48 (7), 1161–1175.

Mastrorocco, N. and Minale, L., 2016. *Information and crime perceptions: evidence from a natural experiment*. CReAM Discussion Paper Series, n. 1601, Department of Economics, University College London.

Meester, W.J., 2004. *Locational preferences of entrepreneurs: stated preferences in The Netherlands and Germany*. Heidelberg: Physica-Verlag.

Meester, W.J. and Pellenbarg, P.H., 2006. The spatial preference map of Dutch entrepreneurs. Subjective ratings of locations, 1983-1993-2003. *Journal of Economic and Social Geography TESG*, 97 (4), 364–376.

Musolino, D., 2015. *Stated locational preferences of entrepreneurs in Italy. The patterns, the characteristics and the explanatory factors of the Italian entrepreneurs' mental maps*. Thesis (PhD), University of Groningen.

Mutinelli, M., 2017. L'attrazione di investimenti diretti esteri, *Milano Produttiva. 27° Rapporto della Camera di Commercio di Milano*, Camera di Commercio di Milano.

Available from: http://www.milomb.camcom.it/documents/10157/34734571/milano-produttiva-2017-parte-prima-capitolo-4.pdf/6199e4ff-8ea5-40c5-b974-2ce84db1ca65 [Accessed 30 October 2017].
Myrdal, G., 1957. *Economic theory and underdeveloped regions*. London: University Paperbacks, Methuen.
OECD, 2006. Milan Italy. *OECD Territorial Reviews*.
Pellenbarg, P.H., 1985. *Bedrijfsrelokatie en ruimtelijke kognitie; onderzoekingen naar bedrijfsverplaatsingprocessen en de subjektieve waardering van vestigingsplaatsen door ondernemers in Nederland*. Sociaal-geografische Reeks 33. Thesis (PhD). Groningen: Geografisch Instituut Rijksuniversiteit Groningen.
Pellenbarg, P., 2012. *Da Mental map van de Nederlandse onderner*. Groningen: Rijksuniversiteit Groningen, Faculteit Ruimtelijke Wetenschappen.
Rositi, M., 1988. L'analisi del contenuto. *In:* M. Rositi and M. Livolsi, eds. *La ricerca nell'industria culturale*. Roma: La Nuova Italia Scientifica, 59–94.
Saarinen, T.F., 1995. Classics in human geography revisited: commentary 2. *Progress in Human Geography*, 19, 107–113.
Siemens-European House Ambrosetti, 2007. *Osservatorio Siemens per migliorare l'attrattività positiva del Sistema Italia*. Available from: http://www.ontit.it/opencms/export/sites/default/ont/it/documenti/archivio/files/ONT_2007-09-01_00115.pdf [Accessed 30 October 2017].
Sletto, B., 2002. Producing space(s), representing landscapes: maps and resource conflicts in Trinidad. *Cultural Geographies*, 9 (4), 389–420.
Smiley, S. L., 2013. Mental maps, segregation, and everyday life in Dar es Salaam, Tanzania. *Journal of Cultural Geography*, 30 (2), 215–244.
Spencer, C. P. and Dixon, J., 1983. Mapping the development of feelings about the city: A longitudinal study of new residents' affective maps. *Transactions of the Institute of British Geographers*, 8, 373–383.
S&W Spiekermann & Wegener, Urban and Regional Research, 2014. *ESPON MATRICES Final Report*. Available from: www.espon.eu
Spilkova, J., 2007. Foreign firms and the perception of regions in the Czech Republic: A statistical examination. *Journal of Economic and Social Geography TESG*, 98 (2), 260–275.
Spilková, J., 2008. Foreign investors and their perceptions of socio-institutional and entrepreneurial environment in the Czech Republic: A pilot study. *Journal of Geography and Regional Planning*, 1 (1), 4–11.
Stern, E. and Krakover, S., 1993. The formation of a composite urban image. *Geographical Analysis*, 25 (2), 130–146.
Stone, P.J., *et al.* 1966. *The general inquirer: A computer approach to content analysis*. Cambridge, MA: The MIT Press.
Svimez, 2011. *Rapporto Svimez 2011 sull'economia del Mezzogiorno*. Bologna: Il Mulino.
Svimez, 2013. *Rapporto Svimez 2013 sull'economia del Mezzogiorno*. Bologna: Il Mulino.
Tuan, Y.F., 1975. Images and mental maps. *Annals of the Association of American Geographers*, 65 (2), 205–212.
Turri, E., 2000. *La megalopoli padana*. Venezia: Marsilio.
Vadjunec, J. M., Schmink, M., and Greiner, A.L., 2011. New Amazonian geographies: emerging identities and landscapes. *Journal of Cultural Geography*, 28 (1), 1–20.
Van den Bosch, H.J.M., 1977. Het subjektieve moment in het ruimtelijk gedrag: poging tot inventarisatie en evaluatie van de "perceptiebenadering" binnen de geografie, *Geografisch Tijdschrift*, 11, 77–97.

CREATIVE MAPPINGS

Some reflections on mental maps

Lars-Erik Edlund

ABSTRACT

The point of departure for this essay is a map drawn in 1963 by the writer's maternal grandfather. It represents the village of Berg, located in northern Sweden, and depicts his activities as a farmer and hunter. But it is also based on grandfather's collective knowledge of the village. In what follows I will examine mental maps of microspaces that reflect what is important to an individual or to the members of a community. One shows how Aivilik Inuits perceive their local environment; another set of urban maps from Los Angeles, California, are based on the views of residents in different areas. The social divides become strikingly apparent on these mental maps. Among the conspicuous features of my grandfather's map are the images he drew to supplement the various geographical locations he laid out. In this respect one might compare medieval *mappae mundi* that is, maps of the world representing compendiums of all things worth knowing. I also consider the appearance of mysterious gaps on grandfather's map, that is, "the silences". Many general perspectives on mental mapping are suggested by a consideration of the map my grandfather drew.

Introduction

These reflections were set in motion by a map that my maternal grandfather made in 1963 of his home village of Berg, located in Ångermanland, in northern Sweden. This map, unfortunately no longer in existence, is reconstructed in Figure 1. It was illustrative of grandfather's activities as a farmer, forestry worker, fisherman, hunter, and berry picker, but it also encompassed the collective knowledge of his village. Although simple in design, the map was rich in pictorial detail, and seeing it again in my mind's eye never fails to conjure up many thoughts and ideas. I would like to convey some of them here.

Avaberg Mountain, an imagined reality

A decade ago, I was reminded of my grandfather's dedicated work on his village map while reading a passage by the late Swedish writer and member

Figure 1. The original of my grandfather's map of his home village of Berg in the parish of Björna in Ångermanland is unfortunately nowhere to be found. I have tried many times without success to reconstruct it from memory. Graphic artist Maria Sundström has kindly worked with me to create the visualization here.

of the Swedish Academy, Torgny Lindgren, in his 2003 novel *Pölsa* (*Hash*, translated by Tom Geddes, 2004). One day, the protagonist of the novel, Manfred Marklund, 107-year-old former newspaper staff writer living in a nursing home, is given a large piece of white paper and colored crayons to make a decoration for his room. He sets to work very deliberately.

> First of all he took his pen and drew the most important lines: the roads and paths and the river and the lake beaches and main dikes. He marked the site of mountains and springs and he wrote in the most significant names. Then he picked up the colored crayons … and used the three green ones to put in a background of forest and bogs and tiny little fields, with here and there some fir scrub and single pines and yellowing birch. He made the water deep blue with touches of black, with a hint of black also in the wettest places, in the swamps and pools and mud.

> … Then he selected the best red, carmine, and put in the human habitation, all the houses he had seen and remembered or that he had heard mentioned and described, and he wrote their names in the same vivid blood-red color: Avabäck, Inreliden, Lillåberg, Kullmyrliden, Ristjöln, Åmträsk, Björknäset, Lakaberg, Finnträsk, Granträskliden, Nyklinten, Kläppmyrliden, Burheden, Morken, Lillsjöliden, Gammbrinken, Ensamheten, Lillvattnet and Granbergsliden. He even restored Matilda Holmström's burned down house. He joined together the houses and villages and isolated crofts and cultivated fields with roads and paths and bridges and footbridges in bright chrome yellow, and drew the boat jetties and mailboxes and the Lycksele bus route across the map with the ochre-colored crayon. In the bottom right-hand corner he indicated with an arrow the road that went to the sanatarium in Hällnäs.
>
> … He sketched Klåvaberg mountain, no longer bare but covered with new forest, yellowy-green and prettily rounded with even slopes on all sides.
>
> … And Oxkall spring and Hömyrbäck brook and Riskläppen hill and Teresias spring were also allocated their individual names and colors.
>
> But right in the center he put Avaberg mountain. He let it rise up out of Gårdbäck moor and Avabäck moor with its rugged slopes to the south and west and steep cliff and sheer faces to north and east. He let mosses and lichens glint on the boulders and bedrock, he placed dead pines and rusty fox traps in their rightful positions and indicated with white dots the sites where prehistoric stone slabs still lay exposed and bare. Along the south side he let the Avabäck stream rage as in the spring thaw, as if it was the Vindel river at the Vormsele falls, splashing its blue-white water over the stones and heather and dwarf birch, even onto the potato field above Eva Marklund's house. He used a fish symbol to denote the little cave where the trout congregated, at the bottom of the hill.
>
> (Lindgren 2002, pp. 156–158; translation by Tom Geddes)

As it turns out, a young woman named Linda, who works at the nursing home and is one of those who look after the former journalist, has been prospecting for gold in Manfred Marklund's home district in her free time. Marklund gives her his hand-drawn map. Some time later she comes to tell him that she has been successful. The story continues now:

> She had tried to find Avaberg mountain hundreds of times. She had gone on foot, she had taken a snow scooter, she had even borrowed a boat and tried to approach from the lake. She had had the green Geographical Survey map in her hands and the Ordnance Survey map and the Yellow Survey map, she had used binoculars and she had forced herself to sit for hours at the planimeter in the Civil Engineer's Department. And she had found all the other mountains, even eskers and hillocks and knolls. But not Avaberg mountain.
>
> Did he still remember, she asked, once drawing her a map on a big white roll of paper?
>
> "Yes, I remember. With the colored crayons."

"That map," she said, "was a miracle! With that, all I had to do was walk straight to it!"

"Yes," he said. "I know. That's where I lived for the greater part of my life."

"Fancy your being able to remember all that!"

"Imagination is my memory," he said modestly.
(Lindgren 2002, pp. 218–219; translation by Tom Geddes)

On the last page of the novel, she once again asks Marklund how he could have drawn such a map from memory. Marklund answers nonchalantly: "Det var bara alldeles detsamma som att göra en mening [It was just exactly the same as constructing a sentence]."[1]

Like his mental map, Marklund's written map conveys the cultural knowledge he has of this geographical area – Avaberg Mountain and its environs. This does not mean that he gives us a snapshot of the area. What we have instead are different epochs and seasons viewed simultaneously. Even hope is infused into the map. This is precisely why Linda can make use of Marklund's drawing in her hunt for the area's secrets. It serves her far better than the National Land Survey map or the official Ordnance Survey map.

Mental maps of microspace

The map that my grandfather drew, and the map that Torgny Lindgren so masterfully describes in his novel, are examples of inner, mental maps. Grandfather's map was a projection of his own individual mental conception. It encapsulated his activities, but also contained the collective knowledge of his village. The abundant amount of information on the map was the result of 80 years of experience and incorporated the stories of many previous generations.

There is another remarkable map that a retired cashier, Erik Pettersson, drew in 1911 to depict the Skultuna Messingbruk Factory (Erixon 1972). Pettersson's map is the result of an impressive feat of memory. As ethnologist Karl-Olov Arnstberg relates (2001, p. 47), after more than half a century had passed, Pettersson still remembered not only every house, but countless details as well. By contrast, the well-known Swedish ethnologist Sigurd Erixon, who had researched the Skultuna factory for decades, remembered almost nothing about it. The difference was simply that while Erixon had studied the factory as a historical phenomenon, Pettersson actually *identified* himself with the local milieu.

I recall that on my grandfather's map the "village side" of the lake, where most of the village's activities took place, seemed disproportionately large in comparison to the Ordnance Survey map – while other parts of the landscape were reproduced closer to scale. This is not surprising if one consider that mental maps reflect what is important to an individual or a community. It is a phenomenon exemplified by Yi-Fu Tuan in his influential book *Topophilia* ([1974] 1990), based on the work of anthropologist Carpenter (1955). Tuan describes how the Aivilik Inuits perceive their local environment on Southampton Island in Hudson Bay, and how they visualize that perception in their maps. Inuit geographical sketches drawn in the 1920s, prior to the Inuits having seen any modern maps of the area, show the Bell Peninsula significantly enlarged. It is given prominence on their mental maps since this is where the Inuits lived.

The Aivilik Inuit maps also indicate gendered differences in how they represent their immediate environs. For Aivilik women, the most important features on their mental maps are the many trading posts and the paths that link them to their home base, as well as the distances to other villages. For men, however, landmarks are located with reference to the coastline, and boundaries generally play a greater role in their mental maps. "Individuals are aware of different aspects of natural terrain because they have cultural meaning", says archaeologist Engelstad (1991, p. 26).

We all carry around inner maps of our environs. Even contemporary urban individuals refer to inner maps, and these may often reflect collective values. In an often-cited study entitled *Differential cognition of urban residents* (1967), Peter Orleans has illustrated this principle with regard to Los Angeles. His research produced synthesized maps showing that informants from the prosperous white district of Westwood generally have a detailed knowledge of their city and its surroundings, while African-American interviewees in Avelon, close to Watts, were significantly less familiar with their area. For them, the landmark thoroughfares were their local streets, while other districts in the city were only vaguely located as "'out-there-somewhere,' with no interstitial information to connect them with the area of detailed knowledge" (Gould and White [1986] 1992, p. 17). However, acquaintance with the city was at its lowest among the small Spanish-speaking minority in the vicinity of Boyle Heights. This group only had an overview of their immediate neighborhood and the bus depot, "the major entrance and exit to their tiny urban world" (p. 17). The social divides in the city leap out at you in these mental maps.

The wealth of images on the maps

Prominent on my grandfather's map were the many images it contained, in addition to symbols for fields, wetlands, and woodlands. Manor houses, cowsheds, and barns were drawn as seen from the front. High hay-drying racks stood next to the cowsheds, even though such racks were long gone. The map was also replete with images of elk that grandfather had killed during a lifetime of hunting – symbolized by stately elk antlers. There were also pictures of fish of all kinds, including a mysterious running one called *skrat-tabborre*,[2] – actually a water salamander – that could climb trees. The map also showed some places where an ominous light called *fegljus*[3] had been observed, and others where people had seen vitter cows, troll cats (a kind of troll ball), and other things belonging to the supernatural world – all drawn in a manner designed to ignite the imagination.

At the time, I thought these were strange illustrations because I had never seen them on the Ordnance Survey maps. Later, I was to learn that the appearance of such drawings on maps had an important role to play in earlier times. The medieval *mappae mundi* (Harley and Woodward 1987, Chap. 18; Simek 1988, 1990, p. 31 ff., 1996, p. 40 ff.; Harrison 1998, p. 83 ff.) often showed the known world as a round, globe-shaped mass through which two bodies of water flowed, forming a "T". The inhabited Earth was divided into three continents: Asia, Africa, and Europe, with the Orient usually placed at the apex of the map. But besides purely geographical information, early maps often had images from mythology and history, and may have included extensive written commentary.

One of the best-known of these mappae mundi, the Hereford map (de Bello 1872, 1954; Harley and Woodward 1987, p. 309), compiled around 1290, contains over one thousand images and texts (Figure 2). It shows a circular Earth with Jesus, surrounded by angels, composing the frame as the judge at the end of time. On the left side, we can see the Northern European part of the world, including a picture of a skier (perhaps in Norway), a bear, and – incongruously – what appears to be a monkey. In the Trans-Saharan region, a world that was foreign to the mediaeval cartographer, cavemen appear, along with a mythological reptilian basilisk, a unicorn, and a number of four-eyed creatures and hermaphrodites. Not far from the northern coastline of the Black Sea one notices a group of *anthropophagi* (cannibals) quite calmly polishing off their meal.

The largest of all known mappae mundi was the Ebstorf map, destroyed in the Allied bombing of Hannover in 1943. Geographer Gunnar Olsson calls it

> a *thesaurus sapientiae* or repository of knowledge: in structure, a topographical depiction of the world, but in content a compendium of everything that is worth knowing. Not a descriptive copy of the world as it is, but a normative interpretation of what it ought to be. (2003, p. 13)

The fascinating imagery of medieval maps and what they communicate with their many figures and texts is intended to provide a picture of God's macrospace. In a summary statement on the mappae mundi, Rudolf Simek writes:

> Like modern maps, medieval maps want to depict reality. However, reality here is not narrowed down to the empirical quality of surveying and projecting, but encompasses much more: a physical abbreviation of the whole reality is aspired to, including both the material and the spiritual world. (1990, p. 58)

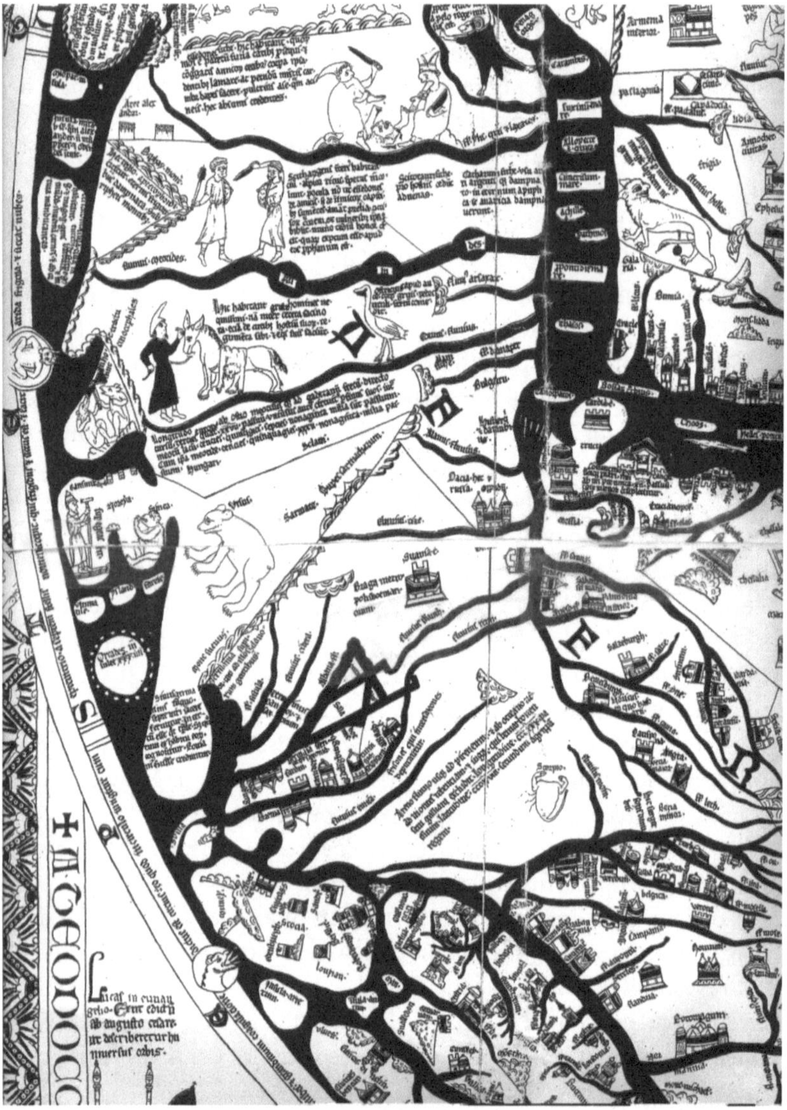

Figure 2. Lower left-hand portion of the Hereford *mappa mundi* (ca. 1290) depicting the Nordic world. Note the skier, bear, and (?) monkey. Above them, a group of cannibals appear to be enjoying a meal (from Ricardus de Bello 1872).

One might say the medieval map was more an expressionistic rendering than a portrait. It also served a different purpose than what we might have expected: it was intended for sinners, not seafarers, as historian and geographer Crosby (1997) has pointed out.

Another well-known map possessing rich imagery is the *Carta Marina* or "Sea map" of Olaus Magnus, originally printed in Venice in 1539 on nine sheets of paper which, when assembled it produced a map 1.70 meters wide by 1.25 meters tall.[4] The Carta Marina could almost be viewed as a pictorial annex to the author's *Opus historia de gentibus septentrionalibus*, published in Rome in 1555. The map of the northern part contains especially vivid imagery, with scenes from various spheres of human activity – hunting, fishing, farming, navigation, trade, worship, etc. It also illustrates a large variety of natural phenomena, in addition to countless animals and fish, some more fantastical than others. In the sea between Iceland and Norway, for example, there appear whales, sea lions, and walruses, as well as many terrifying monsters that presumably dwelt in the depths of the ocean.

To the East of Iceland, the eye is attracted to illustrations depicting mysterious whorls of current. One could easily imagine that Olaus Magnus added them as a kind of decorative motif. However, a pair of oceanographers, Rossby and Miller (2003), discovered that the currents the Carta Marina attempts to depict are remarkably reminiscent of the satellite imagery of this oceanic region, where the warm Gulf Stream encounters colder Arctic flows at the Icelandic-Faroese front. They argue that the "location, size and spacing [of the currents] seem too deliberate to be a purely artistic expression" (p. 87). Could it be that the Carta Marina may be trying to represent the actual currents? We know that Olaus Magnus worked on his map for more than a decade, basing it on historical documents and – perhaps more importantly in this regard – on stories he heard and noted down during his many voyages. He could have gained a detailed knowledge of the currents from the seamen he talked to on the Hanseatic *carracks* (cargo vessels) that traded with Iceland, and whose crews knew those waters extremely well.

> The mariners may have told him about ice conditions, pools of cold water, winds and sea conditions, and about the marine life; and given his evident curiosity he would be one to ask. Perhaps what struck mariners and him the most was the remaining ice along the north coast in Spring and the patches of cold waters, which, we conjecture, he sought to portray as whorls in his Carta Marina (p. 83).

The article cited above attracted considerable attention when it first appeared in the prestigious journals *Nature* (Anon 2004a) and *Science* (Anon 2004b). It suggested that rather than simply dismissing images on old map as artistic decoration, one should at least examine them for whatever truths they may contain. The intention of these maps were to mirror the real world; they were not imaginary in the sense that they showed non-existent regions invented by the cartographer. Instead, they were based as far as possible on data from human observation and experience, and on information derived from books. On the other hand, the unchartered regions did become populated with figments of the imagination, aided by fanciful stories whose origin was hard to trace. While it is doubtful that one may find any of the sea monsters that Olaus Magnus depicts on his map, this has not stopped a number of enthusiasts who, inspired by Rossby and Miller's article, have searched for them in Icelandic waters.

Mental maps: some concluding reflections

As mentioned at the outset, the actual map my grandfather drew of his northern Ångermanland homeland no longer exists in the material world, although it remains very much alive in my memory, and so I have to reckon with the fact that in trying to reconstruct it, much has become blurred over the intervening five decades. As a result, the starting point for this discourse is perhaps may be more accurately stated as *my mental map of my grandfather's mental map*. One thing is certain: my interest in maps was awakened by his hand-drawn sketch on that November day so many years ago, and has remained alive ever since.

Over the years my thoughts have returned many times to that map from my childhood. I am fascinated by the way something in itself so simple brought contemporary and historical perspectives together for me: how the changing seasons are juxtaposed with the activities of daily life that correlated with them, and how the map represented a deep experience of the natural environment as well as the dimensions of inner life bound up with the outside world.

But beyond grandfather's memories, the map told a collective story: it was the villagers' common mental map that grandfather had assembled on one piece of paper, very much like the world depicted in the abovementioned novel *Pölsan* by the former journalist toward the end of his life. In its own way, such a map is a *thesaurus sapientiae*, a richly flowing repository of knowledge, reminiscent of the lively details of the mappae mundi, with the difference that the medieval maps sought to reflect the macrospace and its many mysteries.

Despite the great detail in grandfather's map, as time went on I realized that it contained only what he deemed relevant to illustrate. Historian Jan-Erik Lundström says that maps operate "via the principle of the white lie, by selecting and excluding" (2002, p. 69). Sometimes that which is omitted from a map is what is most revealing – a characteristic the historian of cartography J. B. Harley has called "the map's silences" (2001, p. 86, Magnusson 2002, p. 157; Edney 2005 *passim*).

One of the gaps on grandfather's map especially intrigued me. It was an empty space near the center, so I knew grandfather *must* have been familiar with the location. Then why did he choose not to identify it? Was there an event associated with that place – either something in the village's collective memory, or a matter which concerned him personally – that he was reluctant to talk about? To this day I do not know what it was, but it taught me that maps can also contain *conscious silences*. That, too, was an important lesson I learned, sitting across the kitchen table from my grandfather on that November day 54 years ago.

Notes

1. The Swedish word *mening* could be translated "sentence" as well as "sense", "meaning", "significance".
2. Obviously from vernacular *skratte* "pitiful creature," cf. also Icelandic *skratti* "wizard", "devil", and *abborre* "perch".
3. From vernacular *feg* "death marked" and *ljus* "light".
4. Olaus Magnus and his *Carta Marina* has long attracted the attention of historians and other, see *inter alia*, Ahlenius (1895), Granlund (1951, p. 35 ff.), Richter (1967), Knauer (1981), Johannesson (1982, p. 238 ff.), Balzamo (2005, 2015), and Balzamo and Kaiser (2006).

Disclosure statement

No potential conflict of interest was reported by the author.

ORCID

Lars-Erik Edlund http://orcid.org/0000-0001-8748-9934

References

Ahlenius, K., 1895. *Olaus Magnus och hans framställning af Nordens geografi. Studier i geografins historia*. Uppsala: Almqvist & Wiksell.

Anon, 2004a. News in brief. Map of swirls and sea monsters is spot on. *Nature*, 429, 8–9.

Anon, 2004b. Random samples. Here thar be whorles. *Science*, 304 (5672), 820.

Arnstberg, K.O., 2001. Minnesexpertis. In: M. Elg, et al., eds. *Plats. Landskap. Karta. En vänatlas till Ulf Sporrong*. Stockholm: Stockholm University Press, 46–47.

Balzamo, E., 2005. *Carta marina, 1539. Olaus Magnus édité et raconté par Elena Balzamo*. Collection Merveilleux 26. Paris: J. Corti.

Balzamo, E., 2015. *Den osynlige ärkebiskopen. Essäer om Olaus Magnus*. Stockholm: Atlantis.

Balzamo, E. & Kaiser, R., 2006. *Olaus Magnus. Die Wunder des Nordens. Erschlossen von E. Balzamo & R. Kaiser*. Frankfurt Main: Eichborn.

Carpenter, E.S., 1955. Space concepts of the Aivilik Eskimos. *Explorations*, 5, 131–145.

Crosby, A.W., 1997. *The measure of reality. Quantification and western society, 1250–1600*. Cambridge: Cambridge University Press.
de Bello, Ricardus, 1872. *Hanc quam videtis terrarum orbis tabulam descripsit delineavitque Ricardus de Haldingham sive de Bello dictus. A.S. circa MCCC*. F.T. Havergal, ed. G.C. Haddon, F. Rogers, W. Dutton, artists. Hereford Cathedral. London: Edward Stanford.
de Bello, Ricardus, 1954. *The world map by Richard of Haldingham in Hereford Cathedral circa A.D. 1285*. With memoir by G.R. Crone. (Reproductions of early manuscript maps 3). London: Royal Geographical Society.
Edney, Matthew H., 2005. *The origins and development of J.B. Harley's cartographic theories. Cartographica Monograph 54 = Cartographica 40, nos. 1 & 2*. University of Toronto Press.
Engelstad, E., 1991. The symbolism of everyday life in prehistory. In: E. Baudou, ed., *Archaeology and environment 11* [Report from the 2nd Nordic TAG Conference, Umeå 1987]. Umeå University Press, 23–32.
Erixon, S., 1972. *Skultuna bruks historia. 3. K.-O.* Arnstberg, ed. Stockholms stadsmuseum: Sigurd Erixon kommittén.
Gould, P. and White, R., 1986. *Mental maps*. 2nd ed. (1992). London: Routledge.
Granlund, J., 1951. The Carta Marina of Olaus Magnus. *Imago Mundi*, 8, 35–43.
Harley, J.B., 2001. *New nature of maps. Essays in the history of cartography*. Paul Laxton, ed. Baltimore: Johns Hopkins University Press.
Harley, J.B. and Woodward, D., eds., 1987. *The history of cartography, 1. Cartography in prehistoric, ancient, and medieval Europe and the Mediterranean*. University of Chicago Press.
Harrison, D., 1998. *Skapelsens geografi. Föreställningar om rymd och rum i medeltidens Europa*. Svenska humanistiska förbundet 110. Stockholm: Ordfront.
Johannesson, K., 1982. *Gotisk renässans. Johannes och Olaus Magnus som politiker och historiker*. Stockholm: Almqvist & Wiksell.
Knauer, E.R., 1981. *Die Carta Marina des Olaus Magnus von 1539. Ein kartographisches Meisterwerk und seine Wirkung*. Göttingen: Gratia.
Lindgren, T., 2002. *Pölsan*. Stockholm: Norstedt. [English translation Hash by Tom Geddes. New York: Overlook Press, 2004].
Lundström, J.-E., 2002. Ett pappersark, en planet och det kartografiska begäret. Några exkurser kring världskartans historia. *Glänta*, 3–4, 64–73.
Magnusson, J., 2002. K=A=R=T=O=R. *Glänta*, 3–4, 132–162.
Olsson, G., 2003. Koden till Sanningens ö. *Biblis*, 23, 11–16.
Orleans, P., 1967. *Differential cognition of urban residents. Effect of social scale on mapping. Science, engineering, and the city. A symposium sponsored jointly by the National Academy of Sciences and the National Academy of Engineering*. Publication 1498. Washington, DC.
Richter, H., 1967. *Olaus Magnus Carta Marina 1539*. Lund: Lychnos-biblioteket series 11:2.
Rossby, H.T., and Miller, P., 2003. Ocean eddies in the 1539 Carta Marina by Olaus Magnus. *Oceanography*, 16 (4), 77–88.
Simek, R., 1988. Mappae mundi. *Archiv der Geschichte der Naturwissenschaften*, 22/23/24, 1061–1091.
Simek, R., 1990. *Altnordische Kosmographie. Studien und Quellen zu Weltbild und Weltbeschreibung in Norwegen und Island vom 12. bis zum 14. Jahrhundert*. (Ergänzungsbände zum Reallexikon der Germanischen Altertumskunde 4). Berlin: de Gruyter.

Simek, R., 1996. *Heaven and earth in the middle ages. The physical world before Columbus.* Rochester, NY: Boydell Press. [Translation by Angela *Hall of Erde und Kosmos im Mittelalter. Das Weltbild vor Kolumbus.* Munich: Beck, 1992].

Tuan, Y.-F. [1974] 1990. *Topophilia. A study of environmental perception, attitudes, and values.* New York: Columbia University Press.

Index

Note: Page numbers in *italics* refer to figures and page numbers followed by "n" denote endnotes.